CONTENTS

Introduction: Journey through the Lens: A Friendly Introduction to Microscopy	2
Chapter 1: Decoding Microscopy: Types and Principles Simplified	14
Chapter 2: Light, Compound, and Stereo Microscopes: A Close Look at Microscopy	25
Chapter 3: Electron Microscopes Demystified	37
Chapter 4: Microscope Components	47
Chapter 5: Demystifying Microscopy: Your Essential Guide to Sample Preparation	58
Chapter 6: Mastering Your Microscope, Tips, Tricks and Troubleshooting	69
Chapter 7: Diving Deep into Microscopy, Exploring Specialized Microscopes	80
Chapter 8: Decoding Microscopy, A Beginner's Guide to Capturing and Analyzing Images	91
Chapter 9: Applications of Microscopy	102
Chapter 10: Microscopy Unveiled, Exploring Advanced Techniques	113
Chapter 11: Mastering Microscopy: Troubleshooting Common Issues	124
Conclusion	135

BRIAN RODGERS

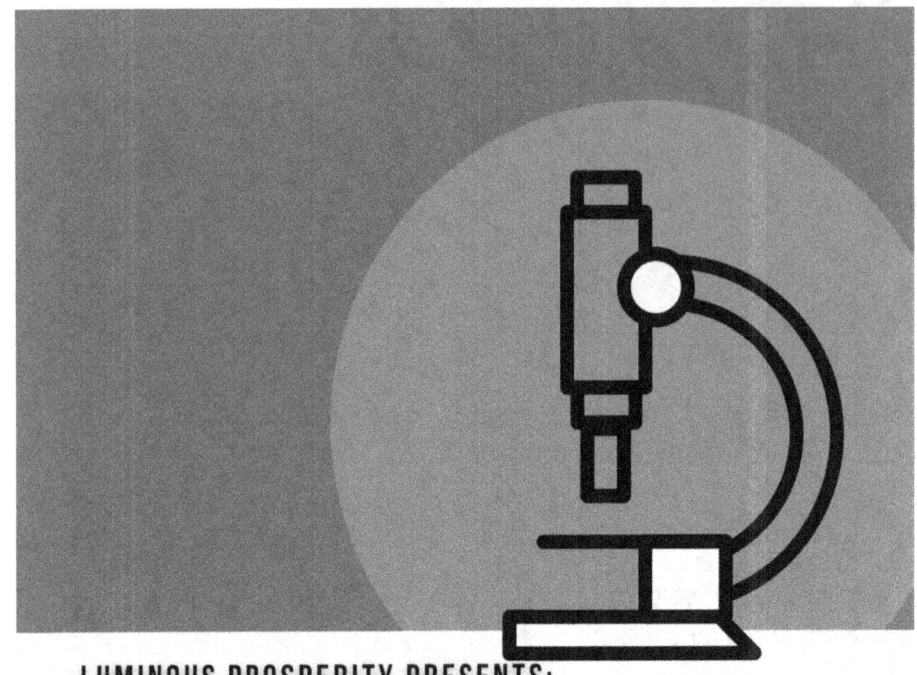

LUMINOUS PROSPERITY PRESENTS:

INTRODUCTION TO MICROSCOPY

WRITTEN BY BRIAN ROGERS

INTRODUCTION: JOURNEY THROUGH THE LENS: A FRIENDLY INTRODUCTION TO MICROSCOPY

Welcome to a fascinating journey through the world of microscopy! We will explore the wonders of this scientific tool that allows us to see the unseen. From its humble beginnings to its modern-day applications, microscopy has revolutionized the way we study the world around us. So, grab your magnifying glass and let's dive into the captivating realm of microscopy.

What Is Microscopy? Understanding the Basics

Microscopy is a fascinating scientific technique that grants us the power to delve into a world beyond the limits of our naked eyes. Imagine being able to see the intricate patterns on a butterfly's wing or the complex structure of a cell — microscopy makes all of this possible and more. It's like having a superpower where the microscopic elements of our universe become visible, understandable, and utterly captivating.

In essence, microscopy involves magnifying objects to such an extent that we can explore their details, textures, and components with incredible precision. This process is achieved through the use of microscopes, instruments designed to reveal the beauty and complexity of the microscopic world. These range from the basic, where visible light is passed through a series of lenses to magnify an object, to the more complex, where electrons or other techniques bring the unseen into view.

This field of study is not just about magnification for the sake of curiosity; it is a crucial pillar in scientific research and discovery. Microscopy lays the groundwork for breakthroughs in various disciplines, including but not limited to, biology, where it enables the study of cells and microorganisms; chemistry, where it assists in observing molecular arrangements; and materials science, where it helps in examining the properties of different substances. Without microscopy, our understanding of these fields would remain surface-level, limiting our potential for innovation and discovery.

Each leap in the development of microscopy has opened up new possibilities, turning the invisible into the visible. It's an ongoing adventure into the minuscule, where each discovery adds another piece to the vast puzzle of understanding our world. Whether it's observing the building blocks of life or examining the fundamental components of the materials around us, microscopy stands as a testament to human curiosity and our relentless pursuit of knowledge.

A Peek into the Past: The Origins of Microscopy

Embark on a time-traveling adventure to the 17th century, where the seeds of microscopy were sown, sprouting a revolution that would forever change how we perceive the microscopic world. It was during this era that the story of microscopy began, marked by curiosity and the desire to explore beyond what the naked eye could see.

The early chapters of microscopy are intertwined with the work of Antonie van Leeuwenhoek, a Dutch tradesman with an insatiable curiosity and remarkable skill in lens crafting. Van Leeuwenhoek, often hailed as the father of microscopy, handcrafted powerful lenses that magnified the unseen with unparalleled clarity. With his simple yet ingenious microscopes, he unveiled a universe teeming with life—discovering bacteria, free-living and parasitic microscopic protists, sperm cells, blood cells, and much more. His meticulous observations laid the foundational stones for the field of microbiology and opened the doors to a world previously beyond humanity's grasp.

But the tale of microscopy doesn't begin and end with Van Leeuwenhoek. Around the same time, another luminary, Robert Hooke, made significant contributions with his groundbreaking book "Micrographia," published in 1665. Hooke's work, featuring detailed drawings of specimens observed under his compound microscope, captured the imagination of the public and the scientific community alike. It was Hooke who coined the term "cell" after observing the structure of cork, setting the stage for cell biology.

These early microscopists, with their rudimentary instruments, were the pioneers on a quest for knowledge, peering into the microscopic abyss and revealing its wonders. Their legacy is a testament to human ingenuity and the relentless pursuit of understanding the fabric of life itself. As we reflect on the origins of microscopy, we honor the spirit of exploration and discovery that propels science forward, reminding us that sometimes, to see the bigger picture, we must focus on the smallest of details.

The Evolution of Microscopes: From Simple to Complex

As we journey further into the story of microscopy, we witness an extraordinary evolution of microscopes, transforming from the rudimentary lenses of the 17th century to the sophisticated tools of today. This progression wasn't just a leap but a meticulous process of innovation and refinement, where each advancement built upon the last, pushing the boundaries of what we can see and understand about the microscopic world.

The transition from simple magnifying glasses to more complex compound microscopes in the early days marked a significant turning point. These early compound microscopes, while a major step forward, were initially limited by optical distortions that blurred the images. The quest for clarity led to innovations in lens design and the introduction of achromatic lenses, reducing these distortions and greatly improving image quality.

But the true revolution came with the leap from optical to electron microscopy in the 20th century. This shift allowed scientists to surpass the limitations of light, using beams of electrons to achieve magnifications up to a million times greater than traditional microscopes. Electron microscopes opened up new dimensions, revealing the intricate details of viruses, the fine structure of metals and materials, and the complex organization of cells at a molecular level.

Further advancements have introduced an array of specialized microscopy techniques, each designed to explore specific aspects of the microscopic world. From the dynamic imaging of living cells with confocal microscopy to the atomic-scale resolution of scanning tunneling microscopes, the evolution of microscopy has equipped scientists with an unparalleled toolkit for discovery.

Each step in the evolution of microscopes has expanded our ability to explore the unknown, enabling scientists to delve deeper into the fabric of the natural world. As we continue to innovate and refine these tools, the journey through the microscopic landscape presses on, with each new generation of microscopes promising to reveal even more of the unseen wonders that await us.

Types of Microscopy: Exploring

Different Methods

Delving into the world of microscopy reveals a fascinating array of techniques, each with its unique way of bringing the unseen into the spotlight. Let's explore some of the diverse methods scientists use to magnify the mysteries of the microscopic world.

Light microscopy, a cornerstone of this realm, uses visible light to illuminate specimens, offering a window into the structure and color of cells and tissues. This method, while fundamental, paves the way for understanding basic biological processes in a wide range of scientific studies.

Stepping into more specialized territory, electron microscopy breaks through the limitations of light to provide incredibly detailed images of ultra-small structures. By harnessing beams of electrons, this technique allows us to peek at the nanoscale intricacies of materials, biological specimens, and even individual atoms, unveiling details invisible to conventional light microscopes.

For those intrigued by the vibrant dance of living cells, fluorescence microscopy is a true spectacle. This method employs fluorescent dyes that glow under specific light wavelengths, highlighting dynamic biological activities within cells and tissues. It's like lighting up a city at night, revealing the bustling activity within.

Meanwhile, scanning probe microscopy stands out for its ability to map surfaces down to the atomic level. It gently 'feels' the surface of a sample with a fine probe, building a detailed topographic map that showcases the bumps and valleys of atoms and molecules.

Confocal microscopy offers another layer of depth, quite literally, by using laser light to scan samples at various depths. This results in clear, three-dimensional images of thick specimens, making it invaluable for detailed structural analysis.

Each of these methods opens a unique door to understanding the fabric of the natural world, from the grand dance of living cells to the intimate embrace of atoms. As we continue to expand our microscopic toolkit, the possibilities for discovery and innovation seem as boundless as the universe itself. So, let's keep our eyes on the lens and our minds open to the endless wonders waiting to be revealed.

The Magic Behind Magnification: How Microscopes Work

Dive into the enchanting mechanics of microscopes, where the magic of magnification brings the microscopic world into focus. At its heart, a microscope transforms the seemingly invisible into a realm of astonishing detail and color, using an elegant interplay of light and lenses.

To start, let's illuminate the pathway of light as it embarks on this transformative journey. Imagine a beam of light directed at a tiny sample. This light, whether from a lamp or a laser, is first passed through a condenser lens, which focuses the light onto the specimen. It's this concentrated illumination that lights up the stage for the microscopic performance to unfold.

Next, the objective lens comes into play, acting as the primary magnifier. This lens collects the light that has interacted with the specimen, bending the rays to magnify the image of the object. Here's where the magic happens: the magnified image is then projected through the tube of the microscope to the ocular lens, or eyepiece, where it undergoes further magnification. The result? A detailed, enlarged view of the specimen that's ripe for exploration.

But the beauty of a microscope doesn't end with magnification. Resolution, the ability to distinguish two close objects as separate, is equally crucial. It's what gives an image its crispness and depth. Advances in optical technology have led to the development of high-resolution lenses and imaging techniques, allowing scientists to observe the fine structure of cells and even individual molecules with astonishing clarity.

The journey of light through a microscope is a testament to human ingenuity, transforming simple rays into a vivid exploration of the microscopic. With each turn of the focus knob, a new layer of detail emerges, inviting us to delve deeper into the wonders of the unseen world.

Unveiling the Invisible: The Importance of Microscopy

Microscopy is more than just a tool for seeing the tiny; it's a vital instrument that has fundamentally transformed our understanding of the world at a microscopic level. This technology has been pivotal in bridging the gap between the known and the unknown, allowing us to explore the intricacies of life, matter, and the universe with remarkable detail and accuracy. By bringing the microscopic into view, microscopy has empowered scientists and researchers across various fields to make observations and discoveries that were once beyond imagination.

In the realm of medical research, for example, the role of microscopy cannot be overstated. It has enabled us to peer into the cellular architecture of the human body, identifying the cellular anomalies that lead to diseases and guiding the development of targeted treatments. In the field of microbiology, microscopy serves as the window through which we observe the behaviors and interactions of microorganisms, enhancing our understanding of their roles in ecosystems, human health, and disease.

Beyond biology and medicine, microscopy has profound applications in materials science, where it aids in the examination of the structural properties of materials at the atomic level. This knowledge drives innovation in creating stronger, more flexible, and more sustainable materials. Additionally, in environmental science, it helps in the analysis of pollutants and microplastics, contributing to efforts in conservation and environmental protection.

The critical insights gained through microscopy underscore its significance in advancing scientific knowledge and technological progress. By enabling us to unveil the invisible, microscopy continues to be a cornerstone of scientific inquiry and discovery, pushing the boundaries of what we know and reshaping our perspective of the microscopic world.

Microscopy in Action: Real-World Applications

The versatility and utility of microscopy stretch far and wide, touching various aspects of science, healthcare, industry, and even our daily lives in ways we might not immediately realize. In the world of medicine, the role of microscopy is particularly striking. It provides doctors and researchers with a critical window into the microscopic realm, allowing for the detailed examination of tissues, the identification of cancerous cells, or the detection of bacterial and viral pathogens. This microscopic insight is indispensable for accurate diagnoses, guiding treatment decisions that save lives.

Microscopy's influence extends beyond the medical field into the arenas of materials science and engineering. Here, scientists employ powerful electron microscopes to probe the atomic structure of materials, uncovering the secrets to their strength, conductivity, or optical properties. This knowledge is crucial in the development of new materials for use in everything from aerospace to electronics, paving the way for innovations such as more efficient solar cells or stronger, lighter alloys.

The reach of microscopy also spans into the realm of environmental science. Researchers utilize microscopy to analyze samples of air, water, and soil, identifying pollutants and tracking their sources. This microscopic evidence is vital for informing conservation efforts, guiding policies to protect our planet, and ensuring the health and safety of ecosystems and communities.

Moreover, in the fields of forensics and archaeology, microscopy serves as a powerful tool for uncovering the truth. Whether analyzing fibers from a crime scene or examining ancient artifacts to unravel the stories of past civilizations, microscopy offers a unique lens through which to observe and interpret the physical traces left behind.

Through these diverse applications, microscopy empowers us to explore, understand, and innovate across the spectrum of human endeavor, illuminating paths forward in science, healthcare, environmental protection, and beyond.

The Future of Microscopy: Innovations on the Horizon

As we look towards the horizon, the landscape of microscopy is shimmering with potential, ready to unfold a new chapter of exploratory marvels. The field is on the cusp of breakthroughs that promise to deepen our insight into the microscopic universe in ways we've only begun to imagine. Innovations like super-resolution microscopy are already dismantling the barriers that once confined our vision, offering a glimpse into the cellular workings at a level of detail previously thought unreachable. Imagine observing the dance of individual molecules within a cell or tracking the intricate processes of life as they happen, in real time and with crystal-clear precision.

Then there's the frontier of cryo-electron microscopy, a technique that freezes biological specimens in their natural state, capturing snapshots of life's building blocks in astonishing detail. This technique is not just enhancing our understanding; it's revolutionizing it, providing a window into the atomic structure of proteins and other biomolecules, and thus paving the way for medical and technological advancements that could transform our world.

But the innovation doesn't stop there. The integration of artificial intelligence and machine learning with microscopy is opening doors to automated analysis and interpretation of images, making the invisible world not only more visible but also more understandable. These tools are set to revolutionize how we approach complex biological questions, making microscopy not just a tool for seeing, but also for discovering and understanding on a scale never before possible.

The future of microscopy is bright with promise, inviting us to peer ever deeper into the fabric of the natural world. With each technological stride, we inch closer to unveiling the mysteries of the universe, one microscopic discovery at a time.

Getting Started with Microscopy: Tips for Beginners

Embarking on the journey of microscopy is like opening a door to a magical dimension, teeming with details and stories waiting to be discovered. If you're eager to start this adventure, here are some friendly pointers to guide you through the basics and help ensure your explorations are both enjoyable and enriching.

First and foremost, immerse yourself in the world of microscopy by reading up on its fundamental principles. Understanding the difference between various types of microscopes, such as compound, electron, and fluorescence microscopes, will help you identify which one aligns with your interests and the type of observations you wish to make.

Choosing the right microscope is a pivotal step in your journey. Consider starting with a simple, user-friendly model that doesn't compromise on quality. Many beginners find compound microscopes to be an ideal choice due to their versatility in viewing a wide range of samples. Don't hesitate to seek advice from seasoned microscopists or educators who can offer insights based on your specific needs and budget.

Preparing your specimens is an art in itself. Take time to learn about different preparation techniques, such as staining, which can enhance the visibility of certain features in your samples. Patience and practice are your best friends here, as mastering sample preparation can significantly elevate your microscopy experience.

Lastly, mastering the adjustments on your microscope, such as focusing and manipulating the light source, is crucial for capturing crisp, clear images. Experiment with these settings to understand how they affect the outcome, and remember, every slide offers a new opportunity to hone your skills.

As you venture into the microscopic world, let your curiosity lead the way. Explore a variety of samples, from the petals of your favorite flower to a drop of pond water, and discover the extraordinary stories they hold. Happy magnifying!

Seeing the World in a New Light: Microscopy and Scientific Discoveries

Microscopy has been the key to unlocking mysteries that have long eluded our grasp, allowing us to venture into the microscopic depths where some of the most pivotal scientific revelations have been found. The tiny cell, once an invisible speck, was brought into the light by Robert Hooke, setting the stage for centuries of biological exploration. Rosalind Franklin's work, revealing the intricate double helix of DNA through X-ray crystallography, provided the framework for modern genetics and a deeper understanding of life itself.

These landmark discoveries share a common thread: the power of microscopy to illuminate the unseen and pave the way for advancements that redefine our understanding of the world.

It's through the lens of a microscope that we've decoded the complexities of viruses, peered into the atom's heart, and traced the pathways of diseases within cells.

As the journey of microscopy continues, it beckons us forward with the promise of uncharted territories and unknown wonders waiting to be discovered. It encourages a fusion of technology and curiosity, where each new advancement not only brings the microscopic world into clearer view but also expands the horizons of human knowledge. The stories of these scientific achievements inspire us to keep looking closer, exploring further, and always seeking to understand the intricate tapestry of life that microscopy reveals.

CHAPTER 1: DECODING MICROSCOPY: TYPES AND PRINCIPLES SIMPLIFIED

Microscopy is a fascinating field that allows us to explore the unseen world at a microscopic level. By using powerful tools and techniques, scientists and researchers can uncover intricate details of cells, tissues, and materials that are not visible to the naked eye. In this chaptert, we will delve into the different types of microscopes and their key principles, simplifying complex concepts to make learning about microscopy accessible and engaging.

The Light Microscope: Your Gateway to the Micro World

The light microscope stands as the initial threshold into the enchanting micro universe. This classic instrument, driven by the simplicity of visible light, magnifies the world of tiny wonders, making it accessible to both novices and seasoned researchers. Imagine peering through its lens to discover a kaleidoscope of cellular patterns and microorganisms, each with their own stories, structures, and secrets. It's this magical reveal, powered by beams of light, that demystifies the microscopic realm lying right beneath our fingertips.

What sets the light microscope apart is its inherent user-friendliness and practicality. It doesn't require complex preparation of samples or the usage of high-tech facilities, making it an ideal starting point for those embarking on their microscopic journey. Through its lenses, students can witness firsthand the bustling activity within a drop of pond water or the intricate designs woven into plant tissues. It's like holding a magnifying glass up to nature, only this glass reveals a world teeming with life and patterns invisible to the naked eye.

The light microscope does more than just enlarge images; it opens doors to understanding the fundamentals of biology and materials science. By illuminating specimens with a spectrum of colors, it lays bare the structural complexities of various samples, offering insights into their form and function. This ability to observe and analyze is crucial for educational exploration, scientific discovery, and even in medical diagnosis, where understanding the microcosm can have profound implications.

In the panorama of microscopy, the light microscope may seem like a humble tool, yet its contribution to science and education is immeasurable. It serves as a beacon, guiding curious minds into the microscopic world, where the small scale does not mean insignificant. As we continue to explore and innovate within this fascinating field, the light microscope remains a cherished instrument, inviting us to look closer and marvel at the wonders it reveals.

Diving Deeper with Electron Microscopy

Electron microscopy, a marvel of modern science, takes us far beyond the limits of traditional light microscopy, into a realm of astonishing detail and precision. This technique leverages the unique properties of electrons, which, due to their minuscule wavelength, can unveil structures at the nanometer scale. Imagine being able to discern the intricate textures of a virus's surface, or the delicate interplay of atoms in a piece of material.

That's the power electron microscopy puts at our fingertips.

Transitioning from light to electrons as the source of illumination transforms the way we perceive the micro world. It's akin to swapping a flashlight for a laser, piercing through layers of complexity to reveal the hidden nuances of microscopic structures. This shift in perspective is not just about achieving higher magnification. It's about unlocking a level of detail so profound that it reshapes our understanding of biology, chemistry, and physics.

The heart of electron microscopy lies in its ability to produce images of unparalleled clarity. By focusing a beam of electrons onto a sample, it captures shadows and contrasts created by the interaction of the electrons with the sample's atoms. This interaction is meticulously recorded, yielding images that can highlight features down to the atomic level.

One of the most compelling aspects of electron microscopy is its versatility. Scanning electron microscopes (SEM) provide a three-dimensional view of the surface of samples, offering insights into texture, morphology, and topographical features. In contrast, transmission electron microscopes (TEM) allow us to peer through ultra-thin slices of specimens, revealing the internal structures with exceptional detail.

Embarking on an exploration with electron microscopy invites us to push the boundaries of visibility and understanding. It's a journey that requires patience and precision, but the rewards are immense. Through this sophisticated lens, the minutiae of the micro world stand in sharp relief, offering clues to some of science's most challenging questions.

Exploring Life with Fluorescence Microscopy

Fluorescence microscopy invites us on an illuminating journey to see the living world in a whole new light—literally. By incorporating the vibrant palette of fluorescent dyes and proteins, this technique allows scientists to paint various components of cells and tissues with glowing markers. As we delve into the microscopic landscape, fluorescence microscopy shines a spotlight on the specific molecules and pathways at work, transforming the invisible into a spectacle of glowing details.

What makes fluorescence microscopy truly captivating is its ability to bring the dynamic dance of life into view. Researchers can track how molecules move, interact, and change within living cells in real-time, providing a front-row seat to the cellular processes that drive life. This method doesn't just highlight the players; it reveals the intricate choreography of biological systems, from the rapid signaling between neurons to the meticulous division of cells.

The versatility of fluorescence microscopy extends its reach across various fields of study. It's a key player in unraveling the complexities of neurobiology, immunology, and developmental biology, among others. Each fluorescent marker is carefully chosen to latch onto a specific target, allowing for the detailed mapping of cellular functions and the architecture of tissues. This specificity opens the door to targeted investigations, whether it's pinpointing the location of a protein implicated in disease or observing the activation of genes during development.

Beyond its scientific prowess, fluorescence microscopy serves as a bridge between art and science. The stunning images it produces are not only valuable for research but are also celebrated for their aesthetic appeal, reminding us of the beauty that lies in the foundational blocks of life. Engaging with fluorescence microscopy is like having a microscope and a paintbrush, offering both a lens to examine the minute details of life and a canvas to highlight the wonder of biological processes in brilliant color.

Getting Up Close with Scanning Probe Microscopy

Venture into the world of scanning probe microscopy (SPM), and you'll find yourself at the frontier of nanoscale exploration. This technique is a testament to human curiosity and ingenuity, allowing us to interact with the surface of materials at an atomic level. Imagine a tiny, precise probe gliding across a sample, not unlike a blind person reading Braille, where each atomic bump and groove tells a story of the material's characteristics and secrets.

Scanning probe microscopy is a collective term for methods that include atomic force microscopy (AFM) and scanning tunneling microscopy (STM), among others. These techniques vary in their approach but share a common goal: to map out the nanoscale topography and properties of materials with astonishing detail. AFM, for instance, feels the surface with a delicate tip attached to a cantilever, measuring the force between the tip and the sample to create an image. Meanwhile, STM measures the tunneling current that flows when the tip is brought very close to the surface, providing insights into the electronic structure of the sample.

What makes SPM stand out is its unparalleled ability to not only image surfaces but also to measure physical properties such as conductivity, magnetism, and mechanical strength at the nanometer scale. These insights are crucial for advancing nanotechnology, material science, and even biology, where understanding the surface properties at such a fine scale can lead to breakthroughs in technology and medicine.

Delving into the world of scanning probe microscopy opens up a landscape where the barriers between the known and unknown blur. Each scan reveals more about the fundamental nature of materials, pushing the boundaries of what we can see and understand. It's an exciting, intricate process that underscores the vast potential waiting to be unlocked in the very fabric of materials, inviting us to keep exploring, discovering, and marveling at the wonders of the nanoworld.

The Versatility of Compound Microscopes

Embarking on a journey through the microscopic world, the compound microscope emerges as a pivotal companion, blending high magnification and sharp resolution to bring the tiniest of details into clear view. These microscopes, renowned for their layered lens system, magnify the marvels of the microcosm, offering a window into the complexities of cellular structures, microorganisms, and the intricate patterns woven into the fabric of materials.

What sets compound microscopes apart is their adaptability across a spectrum of scientific endeavors. They stand at the forefront of educational discovery, where learners of all ages can gaze into the lens and be transported into the cellular workings of plants and animals, sparking curiosity and a profound appreciation for the intricacies of life. In research labs, they become the tools through which new knowledge is forged, from unraveling the mysteries of disease to pioneering advancements in material science.

The design of compound microscopes, featuring an assembly of objective lenses, allows for an exploration of samples at varying levels of magnification. This flexibility is crucial for tailoring the view to specific investigative needs, whether zooming in on the delicate structure of a virus or examining the orderly arrangement of cells in tissue. The ability to adjust magnification not only enhances the depth of study but also fosters a dynamic learning environment, where discovery is limited only by one's curiosity.

Moreover, these microscopes don't just magnify; they illuminate the unseen. Equipped with specialized lighting techniques, such as phase contrast and fluorescence, compound microscopes reveal the hidden stories of transparent and fluorescently tagged specimens, making visible the unseen activities and interactions at play within cells and tissues.

In essence, the compound microscope is more than just an instrument of magnification. It is a gateway to understanding, a tool for education and discovery that transcends disciplines, inviting us to peer deeper into the natural world. Its versatility and capability make it an indispensable asset in both the educational sphere and the cutting edge of scientific research, embodying the spirit of exploration that drives us to learn and understand more about the world around us.

Zooming In with Digital Microscopy

In the realm of microscopy, digital microscopy is like stepping into the future, where the fusion of traditional techniques with digital technology brings a new depth to our exploration of the micro world. This innovative approach harnesses the power of digital cameras and sophisticated imaging software, allowing for the capture, storage, and detailed analysis of specimens in a way that was once beyond our reach.

What truly sets digital microscopy apart is its ability to facilitate not just the observation but also the sharing of discoveries. Scientists can now easily document their findings, creating a visual library that can be revisited and shared with the global research community. This collaboration is vital for advancing our collective knowledge and understanding of the microscopic universe.

Moreover, the integration of digital tools enhances the analytical capabilities of researchers. Through digital microscopy, images can be manipulated to highlight specific features, measure dimensions with astonishing accuracy, and compare changes over time. This level of detail provides invaluable insights, whether for confirming a hypothesis in a research lab or diagnosing a condition in a clinical setting.

Another remarkable benefit of digital microscopy is its contribution to education. With the ability to project images for a group or share them online, educators can engage students in the wonders of the micro world with unprecedented clarity and detail. This accessibility encourages a hands-on approach to learning, making the invisible visible and fostering a sense of wonder and curiosity in the next generation of scientists.

Digital microscopy, therefore, is not just about enhancing the way we see the microscopic world; it's about expanding the possibilities of what we can discover, learn, and share. It invites us to look closer, delve deeper, and connect more broadly than ever before, embracing the vast potential of the microcosm that awaits our curious gaze.

Understanding Stereomicroscopes for a 3D View

Dive into the captivating world of stereomicroscopes and prepare to be amazed by the depth and dimension they bring to the microscopic realm. Unlike traditional microscopes that offer a flat, two-dimensional glimpse, stereomicroscopes, fondly known as dissecting microscopes, unlock a three-dimensional vista, allowing observers to appreciate the true spatial relationships and textures of their specimens. This unique perspective is achieved through the use of two distinct optical paths, one for each eye, mimicking the way our binocular vision works in everyday life.

What makes stereomicroscopes particularly special is their versatility. They are the go-to choice for examining anything that's too large or thick for the confines of a glass slide. Whether it's a biologist gently dissecting a plant to study its internal structures, a forensic expert analyzing trace evidence, or an electronics technician inspecting circuit boards, these microscopes offer an unparalleled view that brings every detail into sharp relief.

Their low magnification combined with a generous working distance means not only can you see the whole picture, but you also have the space to manipulate your subject as needed. It's like having a front-row seat to the minute performances that unfold on the stage of your sample.

Stereomicroscopes open up a whole new dimension of exploration, where the beauty of the microscopic world is not flattened into two dimensions but is presented in its full, lifelike glory. This added depth enhances our understanding and appreciation of the specimen, making it an indispensable tool for a wide range of scientific, educational, and industrial applications. Engaging with a stereomicroscope, one can't help but be drawn into the miniature landscapes and narratives that unfold beneath its lenses, providing a profound connection to the micro world that is both enlightening and inspiring.

The Power of Phase Contrast Microscopy in Viewing Unstained Specimens

Dive into the realm of phase contrast microscopy, and you'll discover a transformative way of viewing the microscopic world. This ingenious method brings to light the subtle, often invisible, details of transparent specimens, from the delicate intricacies of live cells to the fine structure of bacteria, all without the need for dyes or stains. It's like being given a pair of glasses that reveals the unseen beauty of organisms in their most natural and dynamic state.

Phase contrast microscopy stands out because it cleverly manipulates light to enhance the contrast between transparent objects and their surroundings. By converting slight differences in refractive index—essentially, how light bends as it passes through various parts of a cell—into detectable variations in brightness, this technique unveils the complex inner workings of cells and microorganisms. Imagine watching the pulsing movement of a cell's cytoplasm or the serene dance of tiny cilia in real-time; phase contrast microscopy makes these observations not only possible but vivid.

This method is particularly valuable for researchers and medical professionals who study live specimens. The ability to observe cells as they divide, migrate, or interact with their environment opens up new avenues for understanding biological processes, disease mechanisms, and the effects of treatments without altering the samples' natural conditions.

Embracing phase contrast microscopy means stepping into a world where the invisible is made visible, offering a window into the bustling life of cells and microorganisms. It invites you to experience the wonder of discovery, encouraging a deeper appreciation for the intricate dance of life that unfolds at the microscopic level.

Revealing Hidden Worlds with Darkfield Microscopy

Embark on a captivating journey into the heart of microscopic exploration with darkfield microscopy, where the unseen and overlooked come to light in stunning detail. Through the clever manipulation of light, this technique casts a spotlight on the smallest of specimens against a deep, dark backdrop, transforming them into luminous subjects of wonder. By sidelighting samples, darkfield microscopy enhances the contrast of particles, bacteria, and other microscopic entities that would otherwise blend into obscurity under conventional viewing methods.

This unique approach is not just about seeing but experiencing the micro world in a way that captures the imagination and broadens our understanding. It's especially beneficial for visualizing live organisms and delicate structures without the need for chemical staining, preserving their natural state and allowing their true beauty to shine through. From the dynamic swirl of protozoa in a drop of water to the intricate patterns of fine fibers, darkfield microscopy invites us to witness the elegance and complexity of forms that lie hidden in plain sight.

Scientists, educators, and enthusiasts alike find darkfield microscopy an invaluable tool in fields such as microbiology, hematology, and environmental science. It opens up new perspectives on the natural world, encouraging curiosity and discovery in a way that is both accessible and profoundly enriching. Step into the world of darkfield microscopy, and prepare to be amazed by the hidden wonders that await your adventurous spirit.

CHAPTER 2: LIGHT, COMPOUND, AND STEREO MICROSCOPES: A CLOSE LOOK AT MICROSCOPY

A Closer Look at Microscopy Techniques

The journey of microscopy techniques is a fascinating narrative of human ingenuity and the relentless pursuit of knowledge. What began as a curiosity-driven endeavor with the creation of rudimentary magnifying devices has evolved into a sophisticated science that empowers us to unveil the mysteries of the microscopic world. Modern microscopy has transcended the limitations of traditional optical systems, embracing a variety of methodologies each designed to cater to specific research needs and applications.

At the heart of these developments is the light microscope, which, through its simple yet effective design, has illuminated countless discoveries. However, the quest for greater detail and understanding pushed the boundaries further, leading to the creation of compound and stereo microscopes. These instruments, with their enhanced magnifying capabilities and specialized functionalities, opened new vistas in scientific exploration, from the intricacies of cellular biology to the precise requirements of industrial quality control.

But the story doesn't end here. The advent of electron microscopy brought about a paradigm shift, enabling us to magnify specimens to the nanometer scale and beyond. This leap in magnification capability has not just expanded our view but has fundamentally altered our understanding of biological processes and material structures.

Each step forward in microscopy techniques has been driven by a desire to see more clearly, to understand more deeply. As we continue to refine these technologies, we're not just improving our ability to observe but also expanding the horizons of human knowledge. The evolution of microscopy is a testament to our unyielding curiosity and our relentless drive to explore the unseen.

The Essence of Light Microscopy

At its core, light microscopy stands as a testament to the power of simplicity and elegance in scientific exploration. This method, utilizing visible light to reveal the hidden details of our world, bridges the gap between the seen and the unseen. Light microscopes, with their capacity to harness the fundamental properties of light, illuminate specimens in a way that is both profound and accessible. By directing light through lenses, these microscopes magnify the beauty and complexity of biological specimens, from the intricate patterns of cellular structures to the dynamic interplay of tissues.

This type of microscopy serves as a foundational tool in education and research, providing a window into the microscopic realm that is both comprehensive and enlightening. Its utility in laboratories and classrooms around the globe speaks to its versatility and effectiveness in a variety of settings. Whether it's observing the detailed processes of mitosis or exploring the delicate features of microorganisms, light microscopy offers a gateway to understanding that is as intuitive as it is insightful.

What distinguishes light microscopy from other forms, however, is not just its ability to magnify, but its role in nurturing curiosity and discovery. This approach to microscopy does not require complex preparation of the specimen, allowing the natural beauty and intricacy of the sample to stand at the forefront of observation. It's a method that democratizes science, making the exploration of the microscopic world accessible to novice and expert alike.

In the grand tapestry of microscopy, light microscopy is a thread that weaves through the history and future of scientific exploration, embodying the spirit of inquiry that drives us to look closer, dig deeper, and understand more fully. It invites us to marvel at the world at our fingertips, reminding us that sometimes, to see further, we need only to shine a light.

Understanding Compound Microscopes

Diving deeper into the realm of microscopic observation, we encounter the compound microscope—a sophisticated evolution of the light microscope that magnifies the frontier of the unseen. This tool is ingeniously crafted, harnessing a series of lenses to achieve magnifications that bring even the most minute details into sharp relief. It is a gateway to the microcosm, enabling us to peer into the intricate fabric of tissues, cells, and microbes with an unprecedented clarity.

What sets the compound microscope apart is its ability to stack magnification through multiple lenses—namely the objective and the eyepiece lens. This configuration not only amplifies the image of the specimen but also enhances the resolution, allowing scientists and researchers to discern structures separated by infinitesimal distances. The brilliance of this design lies in its precision and depth, making it an indispensable instrument in the arsenal of medical research, microbiology, and the biological sciences.

The compound microscope's prowess is not just in its magnifying capabilities but also in its role as a bridge to the microscopic world. It has propelled advancements in understanding disease pathology, microbial life, and cellular processes, offering a vantage point that is as informative as it is inspiring. Through its lens, the mysteries of life unfold in layers and dimensions that were once beyond our grasp.

In the ever-expanding universe of microscopy, the compound microscope stands as a pillar of optical innovation. It exemplifies how a deeper look, enabled by the right tools, can unravel the complexities of the natural world. With each adjustment of its focus, it not only brings the unseen into view but also draws us closer to the essence of exploration and discovery that defines scientific inquiry.

The Versatile World of Stereo Microscopes

Stereo microscopes, in their unique capacity, offer us a window into the textures and contours of the world in a way that is unmatched by their counterparts. They afford users the luxury of depth perception and a three-dimensional view, making them indispensable tools in an array of applications that stretch from the intricacies of biological research to the exacting demands of precision engineering. Their utility is further enhanced by the feature of a larger working distance, which not only facilitates the manipulation and dissection of specimens but also opens up a landscape of observational opportunities that are otherwise inaccessible with more traditional microscopy techniques.

What truly sets stereo microscopes apart is their ability to bring the observer into a more intimate engagement with the specimen. This is not merely observation—it is an exploration of a world that exists in the spaces and dimensions that our human eyes cannot naturally perceive. In forensic analysis, for example, the stereo microscope becomes the bridge between the visible world and the minute evidence that can tell a deeper story. Similarly, in the realm of archaeology, these microscopes allow researchers to traverse the delicate surfaces of artifacts, revealing the history etched in their forms.

In the domain of electronics assembly, the stereo microscope transcends its role as a tool and becomes an extension of the artisan, enabling the precision and accuracy required in the construction and examination of complex circuits. The benefits here are twofold: enhancing the quality of workmanship and mitigating the potential for error.

Stereo microscopes, with their distinct design and capabilities, enrich the landscape of microscopy. They underscore the diversity of human inquiry and the myriad ways in which we seek to understand and interact with the world around us. Through their lenses, we are reminded of the complexity and beauty of the microscopic realm—a testament to the depth of our curiosity and the breadth of our explorative spirit.

Comparing Light, Compound, and Stereo Microscopes

Navigating the landscape of microscopy reveals a rich tapestry of instruments, each tailored to specific facets of exploration. Light microscopes serve as the venerable pioneers in this field, offering a broad vista for viewing transparent, thin specimens, a testament to the elegance of simplicity in magnification. They illuminate the path for initial forays into the microscopic world, providing a fundamental, yet profound, glimpse into the structures that underpin life and material.

Elevating our gaze brings us to the domain of compound microscopes, the sophisticated descendants of their light microscope ancestors. With their intricate system of multiple lenses, they catapult the observer into realms of higher magnification and resolution, enabling a dive into the cellular and subcellular intricacies that remain elusive to the light microscope. This leap in detail unveils a universe within the minuscule, a closer inspection of life's intricate tapestry.

Venturing further, stereo microscopes offer a different kind of voyage. Unlike their counterparts, these instruments invite us to appreciate the three-dimensional contours of the microscopic landscape. They enhance our perception of depth, granting us the ability to manipulate and examine specimens with an intimacy that flat images cannot convey. This capability is invaluable in a variety of practical applications, from biological dissections to the intricate work of precision engineering, where the interplay of light and shadow reveals the true topography of our tiny subjects.

Together, light, compound, and stereo microscopes form a triumvirate of optical power, each with its own niche in the grand endeavor of exploration. Their unique capabilities and applications underscore the diversity of microscopy, a field that continues to expand the frontiers of our knowledge and understanding.

Breaking Boundaries with Electron Microscopes

As we venture further into the uncharted territories of the microscopic world, electron microscopes emerge as formidable explorers, transcending the limitations imposed by traditional light-based microscopy. These sophisticated instruments, including the scanning electron microscope (SEM) and the transmission electron microscope (TEM), harness the power of electron beams to penetrate the veil of the minuscule, achieving magnifications and resolutions that were once deemed unattainable.

Electron microscopy has not just expanded our vision; it has redefined it. With the SEM, we gain the ability to scrutinize the surface of specimens with unparalleled clarity, revealing the intricate textures and topographies of samples at the nanoscale. The TEM, on the other hand, allows us to plunge into the very essence of materials and biological specimens, unveiling the internal structures with astounding detail and precision. This deep dive into the cellular and subcellular levels uncovers the fundamental building blocks of life and materials, offering insights that are critical for advancing our understanding of disease mechanisms, material properties, and the basic principles of biology.

The journey with electron microscopes is a testament to the boundless curiosity that drives scientific inquiry. It is a journey that not only brings the unseen into light but also challenges our perceptions of the microscopic realm. As we continue to push the boundaries of what can be observed and understood, electron microscopy stands as a beacon of progress, illuminating the path toward new discoveries and innovations that have the potential to revolutionize our world.

Innovations in Microscope Technology

The landscape of microscope technology is undergoing a dynamic transformation, heralding an era of unparalleled clarity and precision in microscopic observation. As the boundaries of traditional microscopy are pushed and prodded by the insatiable curiosity of the scientific community, a wave of innovative microscopes has emerged, each tailored to peel back the layers of the microscopic world with greater fidelity. Among these, the confocal microscope stands out as a beacon of innovation, employing a laser to illuminate the specimen and a pinhole to exclude out-of-focus light. This ingenious configuration results in images of breathtaking detail, allowing researchers to delve into the cellular and molecular intricacies with a newfound sharpness.

The revolution doesn't stop there. Fluorescence microscopy, with its capacity to harness the power of fluorescent dyes, has transformed the way we visualize biological processes. By tagging specific proteins or structures within a cell, this technique illuminates the dance of life under the microscope, providing a dynamic view of the interactions and functions that govern cellular behavior.

Furthermore, the frontier of microscopy is being expanded by nanoscale imaging techniques such as atomic force microscopy (AFM) and scanning tunneling microscopy (STM). These tools extend our vision to the atomic level, enabling the manipulation and observation of individual atoms and molecules. This leap into the nanoscale opens up new realms of potential in materials science and nanotechnology, pushing the envelope of what's possible in the microscopic domain.

In this age of rapid technological advancement, the innovations in microscope technology are not just expanding our view—they are redefining it, opening up new pathways of exploration and discovery in the intricate world that lies beyond the reach of the naked eye.

Improving Image Resolution with Confocal Microscopy

Confocal microscopy emerges as a pivotal innovation, refining the realm of microscopic observation to a degree of precision and clarity previously unimagined. By leveraging a specialized optical design, this technique meticulously focuses a laser light through a pinhole, casting aside any light that doesn't contribute to the focal plane's sharpness. This focused approach yields images of remarkable resolution, allowing for the observation of the minuscule with unmatched detail.

This method shines particularly bright in the fields of biology and neuroscience, where the intricate dance of cellular and molecular interactions unfolds. The ability to cut through the blur of non-essential light means that researchers can peer into the living architecture of cells and tissues, witnessing dynamic processes as they happen. This is not merely observation—it's an invitation into the very pulse of life, offering insights into cellular dynamics that drive understanding forward.

Confocal microscopy's prowess extends beyond the biological, touching upon materials science with the same level of precision. Here, it unveils the complex structures and behaviors of materials at a granular level, enabling advancements in nanotechnology and the development of new materials with unprecedented properties.

The journey into the microscopic, facilitated by confocal microscopy, is a testament to the ceaseless quest for clarity. It represents a convergence of curiosity and technological ingenuity, pushing the boundaries of our visual capabilities and deepening our grasp on the unseen wonders of our world. Through its lens, we are privy to a universe of detail, unfolding in layers and dimensions that challenge the limits of our understanding, driving the spirit of discovery ever forward.

The Power of Fluorescence Microscopy

Fluorescence microscopy stands out as a shining example of how technology can illuminate the depths of the microscopic world in vivid color and detail. This approach employs fluorescent dyes or proteins that attach to specific components of a specimen, bringing them to life under the microscope with a glow that is as informative as it is visually stunning. What sets fluorescence microscopy apart is its ability to zero in on the dance of molecular interactions within cells, providing a dynamic view of biological processes as they unfold.

This technique is a cornerstone in the field of cellular biology, where understanding the location and behavior of molecules in real time is crucial. By tagging proteins with fluorescent markers, scientists can trace the pathways of intracellular processes, uncovering the roles of various proteins in health and disease. The implications of this capability are profound, offering insights into the fundamental mechanisms of life and opening up new avenues for targeted drug development.

Beyond its application in biological research, fluorescence microscopy has a critical role in diagnostics. Its capacity for specificity—able to distinguish between different types of cells and cellular components—makes it an invaluable tool in identifying markers of disease. From the rapid detection of pathogens to the diagnosis of cancer through the identification of biomarkers, fluorescence microscopy enhances our ability to see and understand the complexities of medical science.

In essence, fluorescence microscopy does more than just shed light on the microscopic; it paints a picture of life at the molecular level, enabling us to observe the unobservable. It's a testament to the power of combining scientific curiosity with technological innovation to explore the vast, uncharted territories of the microcosm.

Revolutionizing Microscopy with Nanoscale Imaging

In the realm of scientific exploration, nanoscale imaging techniques like atomic force microscopy (AFM) and scanning tunneling microscopy (STM) have opened new frontiers, allowing us to engage with the world at an almost unimaginably small scale. These powerful methods stand at the vanguard of microscopy, offering a glimpse into the atomic and molecular underpinnings of materials and biological systems. With the capability to visualize and manipulate individual atoms and molecules, they bring us closer than ever to understanding the fundamental forces that shape our universe.

AFM, in particular, operates by scanning a tiny cantilever over the surface of a specimen, meticulously mapping its topography by measuring the force between the tip and the surface.

This technique unveils the surface structures of a wide array of materials with incredible precision, providing insights into their mechanical, electrical, and chemical properties. Meanwhile, STM takes advantage of quantum tunneling to generate high-resolution images of conductive surfaces, revealing the elegance and complexity of atomic arrangements.

Together, AFM and STM challenge our perceptions of the microscopic world, offering tools not just for observation but for manipulation—moving individual atoms and sculpting molecular landscapes. This capacity not only advances our fundamental understanding of physics and chemistry but also catalyzes innovations in electronics, materials science, and medicine. As we continue to push the limits of what can be seen and altered at the nanoscale, these imaging techniques underscore the infinite potential for discovery that lies at the very heart of microscopy.

The Future of Microscopy

The horizon of microscopy stretches far beyond our current vantage point, with emerging technologies promising to redefine our understanding of the microscopic universe. Innovations such as super-resolution microscopy break past the barriers of light diffraction, offering a window into the biological processes at a resolution once deemed unachievable. Meanwhile, the advent of live-cell imaging technologies enables us to witness the dynamic ballet of life in real-time, capturing the ebb and flow of cellular activities without disturbing the natural state of the biological specimen.

These advancements are not just expanding our visual capabilities; they are transforming the way we approach scientific inquiries, allowing for a more nuanced and dynamic exploration of life at the microscopic level. As we forge ahead, the integration of artificial intelligence and machine learning in microscopy promises to unlock even deeper insights, automating complex analyses and revealing patterns previously hidden from human eyes.

In this rapidly evolving field, the future of microscopy is a canvas of potential, inviting us to imagine and explore the unseen with greater clarity, depth, and understanding. It is a journey limited only by our curiosity and the relentless pursuit of knowledge, promising to illuminate the mysteries of the microscopic world with unprecedented detail.

CHAPTER 3: ELECTRON MICROSCOPES DEMYSTIFIED

Unveiling the Power of Electron Microscopes

Electron microscopes stand as titans in the realm of microscopy, offering an unprecedented glimpse into the nanoscopic world that remains elusive to traditional light microscopy. These sophisticated instruments, by harnessing beams of electrons rather than photons of light, unlock a level of detail so profound that the very fabric of materials can be examined at the atomic level. Electron microscopy, in its essence, is a leap into the depths of intricacy that defines our material and biological universes.

The leap from light to electron microscopy is akin to swapping a magnifying glass for a high-powered telescope, where the minutiae of the universe unfold in unexpected clarity and beauty. Electron microscopes achieve their extraordinary resolution by exploiting the shorter wavelength of electrons, allowing scientists to visualize structures that are mere nanometers apart. This transition to electron beams opens up a new dimension of analysis, where the hidden stories of atoms and molecules within cells and materials are revealed.

Electron microscopy, however, is not a singular technique but a gateway to a suite of methodologies, each with its distinct advantages and realms of application. Through the meticulous design of these microscopes, including precise electron optics and advanced detection systems, researchers can tailor their investigative approach to suit their specific scientific quests. Whether it's delving into the internal architecture of a virus with a Transmission Electron Microscope (TEM) or mapping the rugged landscape of a metal alloy with a Scanning Electron Microscope (SEM), electron microscopy equips science with the vision to see the unseen and to ask questions that were once beyond our reach.

In the broader narrative of scientific inquiry, the power of electron microscopes is not merely in their ability to magnify but in their capacity to illuminate pathways to discovery. They serve as a bridge connecting the macroscopic world we navigate daily to the elusive microscopic world, offering insights that fuel innovation and understanding across disciplines. The journey into electron microscopy is a journey into the heart of matter itself, driven by the human quest for knowledge and the relentless pursuit of the truths hidden in the very building blocks of life.

Transmission Electron Microscopes (TEM): A Closer Look

Diving into the intricate universe of Transmission Electron Microscopes (TEM) offers a fascinating perspective on the minuscule. TEMs, with their remarkable ability to transmit electrons through carefully prepared, ultra-thin samples, open windows to the atomic-level intricacies hidden within. This level of exploration transcends what was previously imaginable, providing a lens to witness the dance of atoms and the subtle interplay within materials and biological specimens.

At the core of a TEM's functionality are its meticulously designed components: an electron gun propels electrons into a coherent beam, which electromagnetic lenses then focus and direct. As this electron beam makes its journey through the sample, it interacts with the sample's atoms. These interactions, captured and magnified, materialize on a detector, crafting an image rich with detail about the sample's internal structure. It's a ballet of physics and engineering, enabling us to see beyond the surface and peer into the very essence of matter.

The applications of TEM extend across a broad scientific spectrum. In materials science, it allows for the examination of atomic arrangements that dictate a material's properties. In biology, it provides insights into the organization and function of cellular components, offering clues to understanding life itself. Through TEM, nanoparticles are analyzed to pave the way for advancements in technology and medicine. Each of these applications not only underscores the versatility of TEM but also highlights its critical role in pushing the frontiers of knowledge.

TEM challenges us to meticulously prepare samples and demands a high level of expertise, but the rewards it offers are unparalleled. By magnifying the unseen, it equips us with the knowledge to innovate and discover, continuously expanding the horizons of science and technology.

Understanding the Mechanics of TEM

Delving into the mechanics of Transmission Electron Microscopes (TEM) reveals a fascinating interplay of physics and engineering precision. At the heart of TEM's operation is the electron gun, a component critical for producing a focused beam of electrons with sufficient energy to traverse ultra-thin specimens. This beam, once generated, is meticulously directed towards the sample using electromagnetic lenses. These lenses serve a dual purpose: they not only focus the electron beam but also adjust its trajectory, ensuring that it interacts with the sample in a controlled manner.

As the electrons engage with the specimen, they undergo scattering and diffraction based on the atomic structure they encounter. This interaction is key to TEM's remarkable ability to illuminate the internal composition of materials at the atomic or molecular level. The scattered electrons are then captured, focused, and transformed into a magnified image by a series of electromagnetic lenses positioned downstream from the sample.

The fluorescent screen or digital detector at the end of this journey plays a crucial role in visualizing the final image. It captures the transmitted electrons and translates them into a form that can be observed and analyzed by researchers. This conversion from scattered electrons to a visible image is where the unseen world becomes observable, offering insights into the structural and compositional nuances of the sample under investigation.

TEM's intricate mechanics, from electron beam generation to image formation, exemplify the sophisticated engineering and scientific principles that enable us to probe the microscopic universe with unprecedented clarity and depth.

The Versatile Applications of TEM in Research

TEM's utility in the scientific community is as broad as it is profound, serving as a pivotal tool across an array of research domains. Its capacity to peer into the atomic and molecular architecture of materials has rendered it indispensable in materials science. Here, TEM aids in deciphering the structural underpinnings that govern material properties, paving the way for innovations in everything from semiconductors to biomaterials.

In the realm of biology, TEM opens a window to the cellular and subcellular levels, offering unparalleled insights into the organization, function, and dynamics of biological systems. This deep dive into the biological microcosm has profound implications, enhancing our understanding of disease mechanisms, drug interactions, and the fundamental processes of life itself.

Nanotechnology, a field at the forefront of modern science and engineering, also heavily relies on TEM. The technique's ability to analyze nanoparticles and nanomaterials is critical for the development of new nanoscale devices and materials with novel properties. Whether it's optimizing catalysts for energy applications or designing drug delivery systems, TEM stands at the heart of nanotechnology research.

Beyond these, TEM's influence extends to environmental science, where it helps in identifying particulate matter and pollutants at the microscopic level, and in geology, where it aids in the study of mineral structures and compositions. Its versatility even reaches into the realms of chemistry and physics, where understanding atomic arrangements and bonding can lead to breakthroughs in quantum computing and materials science.

TEM's versatility in research is not just about expanding our knowledge in these domains but also about interlinking them, fostering interdisciplinary collaborations that drive innovation and discovery.

Scanning Electron Microscopes (SEM): Surface Explorers

Scanning electron microscopes (SEM) take us on an exploratory journey across the variegated landscapes of sample surfaces, offering insights with exceptional depth and clarity. These sophisticated instruments employ a finely focused beam of electrons that sweeps across the specimen, a technique that illuminates the surface's complexities through the collection of emitted secondary electrons. This interaction between the electron beam and the sample surfaces brings to light an array of textures, compositions, and topographical details that were previously obscured to the human eye.

The brilliance of SEM lies in its capacity to render these details with remarkable resolution, painting a three-dimensional picture of the sample that is both informative and visually compelling. This capability is underpinned by a precise orchestration of components including an electron gun, electromagnetic lenses, and a scanning system, all working in concert to navigate the electron beam across the specimen with microscopic precision.

As the beam traverses the sample, it unveils features with astonishing clarity, from the rugged terrain of a mineral sample to the intricate patterns on a semiconductor. This high-resolution surface imaging is not just about capturing images; it's about unlocking the stories that these surfaces tell about their composition, functionality, and interaction with their environment.

SEM's contribution to science and research is vast, touching everything from the development of new materials with enhanced properties to the forensic analysis of trace evidence. By allowing scientists to conduct a detailed survey of surfaces down to the nanometer scale, SEM plays a critical role in advancing our understanding of the material world. This journey of discovery, facilitated by SEM, continues to open new frontitudes in scientific inquiry, reflecting the relentless human quest to explore and understand the complexity of the natural world around us.

Delving into the Technical Workings of SEM

At the core of a Scanning Electron Microscope's functionality lies a symphony of precision engineering and sophisticated physics. The initial stage in SEM's operation is marked by the electron gun, a pivotal component responsible for generating a beam of electrons. These electrons, akin to the scouts of the microscopic realm, are accelerated to high energies and focused into a narrow stream. This stream is then expertly steered across the sample's landscape by electromagnetic lenses and scanning coils, which manipulate the beam's path with meticulous control.

As the electron beam dances across the sample surface, it engages in a series of interactions with the atoms it encounters. This interaction leads to the emission of secondary electrons, each a messenger carrying detailed information about the surface's topography and composition. The magic happens when these secondary electrons are captured by a sophisticated detection system. Here, they are converted into a signal that, through a process of amplification and processing, blossoms into a detailed image. This image, rich in texture and depth, unveils the intricacies of the sample's surface with stunning clarity.

The journey from electron beam generation to image creation in SEM is not just a display of technical prowess but a testament to the relentless human pursuit of understanding the microscopic world. By translating the interactions of electrons into visual narratives, SEM provides a unique perspective on the materials and structures that shape our world. This powerful blend of technology and science allows us to explore the microcosm's complexities with an unprecedented level of detail, opening new avenues for discovery and insight across a multitude of scientific fields.

SEM's Impactful Role Across Disciplines

Scanning Electron Microscopy (SEM) transcends its role as a mere instrument, evolving into a linchpin of scientific inquiry across a kaleidoscope of disciplines. Within the realms of materials science, SEM's aptitude for rendering the textures and structures of materials at the nanoscale not only enlightens researchers about the physical properties of substances but also steers the development of innovative materials designed to tackle contemporary challenges. These explorations into the minutiae of materials underpin advancements in industries ranging from aerospace engineering to renewable energy, illustrating the far-reaching implications of SEM technology.

In geology, SEM serves as a powerful tool for deciphering the stories etched into the very fabric of our planet. By magnifying the details of mineral structures and compositions,

SEM aids geologists in understanding the processes that have shaped the Earth over millennia. This microscopic journey through geological samples unveils clues about the planet's past, informing predictions about its future.

The forensic field benefits immensely from SEM's capabilities, as well. The detailed surface analysis it provides enables forensic scientists to uncover microscopic evidence that might otherwise remain hidden. This evidence plays a pivotal role in solving crimes, demonstrating SEM's crucial contribution to societal well-being.

Across each of these disciplines, SEM acts not just as a microscope but as a bridge, connecting nuanced surface details to broader scientific narratives. Through its ability to reveal the unseen with remarkable clarity, SEM empowers scientists and researchers to forge new paths of discovery, illustrating the enduring impact of microscopy on our quest to understand the complex tapestry of the natural world.

Comparing and Contrasting SEM and TEM

Diving into the distinctive realms of Scanning Electron Microscopy (SEM) and Transmission Electron Microscopy (TEM), we embark on a journey to unravel the nuanced differences that set these two microscopy giants apart. At their core, SEM and TEM are both adept at unveiling the microscopic mysteries of the material world, yet they do so from fundamentally different perspectives. SEM specializes in surveying the intricate landscapes of sample surfaces, rendering high-resolution images that reveal topographical and compositional nuances. Its approach is akin to an aerial view, providing a detailed map of the surface terrain.

Conversely, TEM invites us into the inner sanctum of materials, offering a passage through the very fabric of samples. It transcends the superficial, delving into the atomic and molecular intricacies that lie beneath the surface. Where SEM sketches the exterior, TEM elucidates the internal architecture, demanding that samples be sliced into ultra-thin sections to allow the passage of electrons. This requirement for sample preparation reflects one of the stark contrasts between the two techniques, underscoring the meticulous care with which scientists must approach TEM analysis.

Furthermore, the choice between SEM and TEM often hinges on the specific scientific question at hand. Are we seeking to understand the ruggedness of a surface or the organization of internal structures? The decision is not merely technical but philosophical, reflecting the depth and direction of our scientific curiosity. In essence, SEM and TEM are complementary lenses through which we explore the vastness of the microscopic world, each with its own unique strengths and limitations, yet together, they form a comprehensive toolkit for the modern researcher.

Challenges and Limitations of SEM and TEM

Navigating the world of electron microscopy, both SEM and TEM present a series of hurdles that underscore the nuanced complexity of these powerful tools. One of the most formidable challenges lies in the intricate art of sample preparation—a process that demands not only precision but also a deep understanding of the material under investigation. Particularly for TEM, where samples must be sliced to near atomic thicknesses, the preparation is as much an art as it is a science, requiring meticulous attention to detail and patience.

Moreover, both SEM and TEM are susceptible to various artifacts that can mar the clarity and accuracy of the resulting images. These can stem from sample contamination, a common adversary in the microscopic arena, where even the slightest impurity can skew results. Additionally, charging effects—where accumulated electrons alter the surface properties of the sample—pose a significant challenge, especially for non-conductive materials, potentially leading to distorted or misleading imagery.

These limitations, while noteworthy, do not diminish the profound impact of SEM and TEM in pushing forward the frontiers of science. Rather, they highlight the ongoing dialogue between scientific inquiry and technological refinement, a testament to the resilience and ingenuity of the research community in its quest to unravel the mysteries of the microscopic world.

The Future of Microscopy: Evolving Techniques and Technologies

As we peer into the horizon of microscopy, the landscape is abuzz with innovation, heralding a new era of scientific discovery. Emerging techniques, such as super-resolution microscopy, shatter the conventional limits, enabling us to visualize structures at resolutions once deemed impossible. Meanwhile, the advent of correlative microscopy intertwines various imaging methods, offering a multifaceted view of samples that enriches our understanding in unprecedented ways. These advancements are not just expanding our visual capabilities but are also crafting new pathways for analysis, challenging us to rethink the paradigms of microscopic investigation. With each technological stride, we edge closer to unraveling the complex tapestry of life and matter at their most fundamental level. The journey ahead in microscopy is illuminated by the promise of techniques that will further blur the lines between the seen and the unseen, driving forward the ceaseless quest for knowledge that defines the spirit of scientific exploration.

CHAPTER 4: MICROSCOPE COMPONENTS

Optical Components – The Eye of the Microscope

Diving into the realm of optical components, we embark on an intriguing journey through the heart of microscopy. These elements are the microscope's vision, intricately designed to capture and refine light, bringing the microscopic world into vivid detail. At the core of this optical odyssey are the lenses, light sources, condensers, and diaphragms—each playing a pivotal role in the transformation of invisible to visible.

Lenses stand as the cornerstone of magnification. They are tasked with a critical job: bending light rays to bring the minuscule into view. As we delve deeper, we uncover two heroes in the world of lenses—the objective and the eyepiece. The objective lens, positioned near the specimen, acts as the initial magnifier, capturing the scene. Its partner, the eyepiece lens, takes this magnified image and escalates it further, enabling us to peer into the cellular dance of life with astonishing clarity.

Illumination is the stage on which our microscopic performers are revealed. Light sources, therefore, become our spotlight, shedding light on the otherwise invisible. From the warm glow of halogen to the precise beams of LED lights, and the unique glow of fluorescence lamps, each light source has its role, influencing not only visibility but the very quality of the image we observe. It's a delicate balance of brightness and contrast, setting the stage for discovery.

Enter the unsung heroes: condensers and diaphragms. These components do not bask in the glory of producing the image, but without them, our vision would be overwhelmed by chaos. The condenser, with its ability to focus light onto the specimen, works in harmony with the diaphragm, which regulates this light's intensity. Together, they refine the illumination, crafting an environment where each beam of light serves a purpose, enhancing contrast and detail without succumbing to the pitfalls of overexposure or glare.

Through this exploration of optical components, we've ventured into the essence of what makes microscopy not just possible, but powerful. These elements, the eyes of the microscope, do more than just see—they illuminate, magnify, and refine our view of the microscopic universe, allowing us to explore realms beyond the reach of the unaided eye.

Lenses and Their Functions – Magnifying the Minuscule

Lenses are truly the superheroes of the microscopy world, wielding the power to magnify the minuscule and unveil the hidden wonders of the microscopic realm. These vital components act as the microscope's eyes, intricately designed to capture, bend, and focus light rays in a way that enlarges the image of the specimen being studied. Each lens within the microscope plays a specific role in this magnification process, ensuring that we can explore the intricacies of tiny worlds with unparalleled clarity and detail.

In the grand scheme of magnification, two types of lenses stand out for their critical roles: the objective lenses and the eyepiece lenses. The objective lens is the microscope's primary magnifier. Located just a whisper away from the specimen, it gathers light reflected from or transmitted through the specimen, creating an enlarged image of the object. This initial magnification is crucial, as it sets the stage for what comes next.

Following the objective lens's lead, the eyepiece lens, also known as the ocular lens, takes the baton and further magnifies the image. Sitting at the top of the microscope, the eyepiece lens allows the viewer to observe the magnified specimen with ease, offering a window to the microscopic world that is both accessible and rich in detail. The combined powers of the objective and eyepiece lenses enable a deep dive into the microscopic universe, revealing details that would otherwise remain unseen by the naked eye.

Beyond their primary task of magnification, lenses also contribute to the overall quality of the observed image. High-quality lenses can significantly reduce aberrations, such as blurring or color distortion, ensuring that the image remains sharp and true to life across the magnification range. The choice of lens material, curvature, and coating all play a part in this, highlighting the meticulous engineering behind each microscope lens.

The journey of magnification is a collaborative effort between these lenses, a delicate dance of light and optics that brings the smallest details into focus. Through their functions, lenses not only magnify but also enrich our understanding of the microscopic world, inviting us to look closer and discover the beauty hidden in the minuscule. As we continue to explore the vast possibilities offered by microscopy, the role of lenses as the magnifiers of the minuscule remains central, a testament to their invaluable contribution to the pursuit of knowledge and discovery.

Light Sources – Shedding

Light on the Invisible

In the mesmerizing journey of microscopy, light sources play a pivotal role, acting as the beacon that illuminates the path to discovery. These vital components do more than just light up the stage; they transform the invisible into the visible, allowing us to gaze into the microscopic world with clarity and wonder. The selection of a light source is not merely a matter of brightness but involves a nuanced choice that affects the quality and nuances of the image observed.

Halogen bulbs, known for their warm, consistent glow, have long been a staple in microscopy. They provide a level of illumination that brings out the intricate details of a specimen, highlighting features that might otherwise remain hidden in the shadows. However, the evolution of technology has introduced more contenders to the field, each with its unique advantages.

LED lights have emerged as a popular choice, celebrated for their efficiency and longevity. Their cool, bright light offers excellent visibility while minimizing the heat exposure that can alter or damage delicate specimens. LEDs also allow for precise control over the intensity and duration of illumination, a crucial factor when examining sensitive or live samples. Their energy efficiency and minimal heat output make them a sustainable and safer choice for prolonged research sessions.

Fluorescence lamps, on the other hand, unlock a new dimension of microscopic exploration. They do not merely illuminate; they excite. By emitting specific wavelengths of light, fluorescence lamps cause certain materials within the specimen to glow, revealing hidden structures and compositions in a kaleidoscope of color. This ability to highlight biological and chemical markers has revolutionized fields such as microbiology and genetics, offering insights that were once beyond reach.

The choice of light source is a critical decision in microscopy, one that influences not just visibility but the very way we perceive and understand the microscopic landscape. Each light source, from the traditional glow of halogen to the precise illumination of LED and the vibrant revelations of fluorescence lamps, offers a unique lens through which to view the microcosm. By selecting the appropriate light source, we wield the power to shed light on the invisible, embarking on a journey of discovery that illuminates the unseen wonders of our world. As we navigate through the spectrum of options, we are reminded that in the realm of microscopy, light is not just a tool but a gateway to exploration and understanding, inviting us to see beyond the boundaries of the visible and venture into the microscopic frontier with eager curiosity and an open mind.

Condensers and Diaphragms – Controlling the Illumination

Peering deeper into the intricate mechanics of microscopy, we find ourselves exploring the realm of condensers and diaphragms —a space where the control of light transforms from a mere possibility to a finely tuned reality. These elements may not capture the spotlight like lenses or light sources, but their role in the art and science of microscopy is undeniably pivotal.

Condensers, with their ability to focus light directly onto the specimen, act as the guiding hands of illumination. Imagine trying to highlight the delicate features of a tiny organism or the complex structure of a cellular component without the precise direction of light; it would be akin to exploring a vast cave without a flashlight. This is where the condenser steps in, harnessing and directing the light to ensure that every detail is bathed in clarity and contrast. The magic of the condenser lies in its capacity to concentrate light, making the unseen world not just visible but vividly detailed.

Diaphragms play an equally critical role in this dance of light. If the condenser is the hand that guides the light, the diaphragm is the gatekeeper, controlling the intensity and spread of illumination that reaches the specimen. By adjusting the diaphragm, one can fine-tune the amount of light that permeates the microscopic stage, preventing the harsh glare of overexposure and preserving the delicate balance of light and shadow. This meticulous control is crucial for enhancing the depth and detail of the image, allowing for a more nuanced observation of the specimen's structure and behavior.

Together, condensers and diaphragms form a duo of precision and control, enabling microscopists to tailor the illumination to meet the exact needs of each specimen. Whether it's increasing contrast to highlight a particular feature or adjusting the depth of field to focus on a specific layer, these components offer the tools needed to customize the viewing experience. It's through this thoughtful manipulation of light that microscopy transcends simple observation and becomes a gateway to discovery, allowing us to unveil the complexities of the microscopic world with remarkable clarity and depth.

In the grand narrative of microscopy, where every component plays a role in unveiling the hidden wonders of the microcosm, condensers and diaphragms stand out as the masters of illumination. By controlling the flow and focus of light, they not only enhance visibility but also enrich our understanding, inviting us to look closer and delve deeper into the microscopic universe with curiosity and precision.

Mechanical Components – The Microscope's Framework

As we pivot from the world of optics to the realm of structure within microscopy, we find ourselves exploring the unsung heroes that constitute the microscope's skeleton: the mechanical components. These elements, while not directly involved in the magnification and visualization of specimens, are fundamental in providing the stability and precision necessary for any microscopic exploration.

The stage and its accompanying stage clips serve as the platform and anchor for the specimens under study. Picture the stage as a vast ocean where microscopic entities reside; the stage clips act as the lighthouses, guiding and securing these entities in the tumultuous waters of exploration. This setup is critical as it ensures that our microscopic subjects remain centered and stationary, enabling us to delve into their world with unparalleled accuracy and without the distraction of movement or displacement.

Moving to the focus knobs, these components introduce us to the fine art of clarity and detail. The coarse focus knob invites us into the microscopic realm, providing a general overview of the specimen's landscape. As we crave more detail and depth, the fine focus knob steps in like a skilled artist, fine-tuning our view with delicate precision. This dual-knob system empowers us to navigate the microscopic depths with ease, ensuring every journey is both enlightening and fruitful.

The stand and base of the microscope do more than just support the physical weight of the instrument; they are the bedrock upon which all microscopic inquiry is built. Imagine a towering structure; its strength and stability lie in its foundation. Similarly, the stand and base ensure the microscope remains steadfast and durable, allowing for seamless operation and preventing any vibrations or movements that could disrupt our microscopic explorations.

Together, these mechanical components create a symphony of stability and precision, allowing the optical elements to perform their functions to the fullest. They remind us that in the quest for understanding the microscopic, every part of the microscope, no matter how seemingly inconsequential, plays a vital role in unveiling the wonders of the unseen world.

Stages and Stage Clips – Holding the Specimen in Place

INTRODUCTION TO MICROSCOPY

In the grand orchestra of microscopy, where each component plays its pivotal role in unveiling the miniature marvels of our world, the stage and stage clips emerge as the steadfast guardians of stability. It's here, on the stage, that the specimens find their spotlight, poised and ready for their moment under the microscope's scrutinizing eye. This flat platform is not just a surface but a critical arena where the microscopic drama unfolds, held in place by the ever-reliable stage clips.

These clips are more than mere holders; they are the unsung heroes ensuring that our microscopic subjects remain anchored, undisturbed by external tremors or inadvertent nudges. Imagine attempting to delve into the intricate details of a specimen, only to have it drift out of view. Frustration would ensue, clarity would diminish, and the quest for knowledge would be hampered. This is the chaos from which stage clips save us, gripping the specimens with precision and allowing for a seamless exploration of their hidden intricacies.

The dance between the stage and stage clips is a delicate balance of security and accessibility. As specimens are carefully positioned on the stage, the clips gently but firmly secure them, ensuring that every detail remains in clear focus. This partnership allows researchers and hobbyists alike to traverse the microscopic landscape with confidence, knowing that their subjects are precisely where they need to be.

Moreover, the design of stages and stage clips often reflects a keen understanding of user needs, offering versatility for different types of specimens and experiments. Whether it's a slide of pond water teeming with microorganisms or a thin section of rock revealing Earth's ancient stories, the stage and its clips adapt, providing a reliable foundation for exploration.

In this crucial aspect of microscopy, stages and stage clips do not just hold specimens in place; they create the conditions necessary for discovery. They remind us that in the pursuit of the microscopic, every element of our equipment plays a role in shaping our journey, guiding us closer to the truths hidden in the minuscule.

Focus Knobs (Coarse and Fine) – Bringing the World into Focus

Navigating the intricate landscape of the microscopic world requires more than just a keen eye; it demands precision, patience, and the right tools to bring the unseen into sharp relief. At the heart of this quest for clarity are the microscope's focus knobs—coarse and fine. These knobs are not mere adjustments; they are the keys to unlocking a universe of detail, texture, and wonder that lies beyond the reach of the naked eye.

The journey into the microscopic begins with the turn of the coarse focus knob. This initial adjustment is like opening the door to a hidden garden, where the outlines of our subject begin to emerge from the blur. It's a broad stroke, setting the stage for the delicate work that follows. Think of it as the first layer of a masterpiece painting, necessary but incomplete without the subsequent layers of detail and color that give the work its depth and life.

Once the coarse focus has brought the specimen into the general vicinity of clarity, the fine focus knob takes over, acting as the microscope's precision instrument. This is where the magic happens. A gentle twist here, a slight adjustment there, and the details of our microscopic landscape come into sharp focus. This knob is like the artist's brush, adding the final touches that transform a simple sketch into a vivid portrait of the microscopic world.

The dance between the coarse and fine focus knobs is a delicate one. It requires a careful balance, a light touch, and an understanding that the journey from blur to clarity is as much about patience as it is about technique. Each specimen, with its unique structure and composition, presents a new challenge, a new opportunity to explore the depths of the microscopic universe with precision and grace.

In the realm of microscopy, the focus knobs are more than mechanical components; they are the tools through which we navigate the tiny wonders of our world. With each turn, they offer us a clearer view, a closer look, and a deeper appreciation for the complexities and beauties that lie hidden in the minuscule. Through their guidance, the microscopic world comes into focus, revealing secrets and stories waiting to be discovered.

Stands and Bases – The Microscope's Foundation

In the fascinating world of microscopy, the unsung pillars of strength and stability are found in the microscope's stands and bases. These fundamental components are the bedrock on which the microscope rests, crucial for ensuring a steady and unwavering gaze into the microscopic realm. Constructed from durable materials, stands provide the skeleton that supports the intricate optical and mechanical parts, allowing the microscope to maintain its integrity even during intense scientific exploration.

Bases, on the other hand, act as the anchoring platform, grounding the microscope and safeguarding against any disturbances that might disrupt the clarity of observation. Their solid construction is not only about physical support but also about precision. A stable base ensures that vibrations and movements from the external environment are minimized, thus preserving the delicate focus on the specimen being examined.

Imagine a bridge without a sturdy foundation; it would be susceptible to the slightest tremors. Similarly, a microscope without a robust stand and base would struggle to provide the precision needed for clear, detailed observations. It's this foundation that allows for the delicate adjustments and fine-tuning required to explore the complexities of the microscopic world with confidence.

Moreover, the design of these components often reflects a balance between functionality and ergonomics, ensuring that scientists and hobbyists alike can engage in prolonged research without compromising on comfort or stability. As we delve into the microscopic, the stands and bases remind us of the importance of a solid foundation in the pursuit of discovery. They might not capture the limelight as the lenses or light sources do, but their role is indispensable, providing the stability that paves the way for groundbreaking insights and revelations in the vast universe of the minuscule.

CHAPTER 5: DEMYSTIFYING MICROSCOPY: YOUR ESSENTIAL GUIDE TO SAMPLE PREPARATION

Microscopy is a fascinating tool that allows us to explore the intricate details of the world around us at a microscopic level. From studying the structure of cells to analyzing the composition of materials, microscopy plays a crucial role in various scientific disciplines. However, before we can delve into the microscopic realm, we must first ensure that our samples are properly prepared. In this essential guide to sample preparation in microscopy, we will walk you through the steps to effectively prepare your samples for optimal imaging and analysis.

Preparing Samples - The First Step in Microscopy

Embarking on your microscopy adventure begins with the crucial step of preparing your samples. This stage is more than just a prerequisite; it's the foundation upon which all your future observations and discoveries will rest. Imagine trying to read a book with the pages all jumbled up – that's what diving into microscopy without properly prepared samples would be like. Thus, ensuring your samples are well-prepared is essential for a clear, insightful view under the microscope.

The art of sample preparation is varied and can range from simple procedures to more intricate techniques depending on the nature of your specimen and the objectives of your study. To start, you might find yourself cleaning the samples to remove any debris or contaminants that could obscure your view. This might seem mundane, but it's akin to setting the stage before a performance, ensuring that nothing detracts from the specimen's inherent beauty and complexity.

Next comes the drying process for those samples that require it, a careful balance to remove moisture without damaging the delicate structures you wish to examine. Think of it as preparing a delicate dessert that needs just the right amount of time in the oven. Too little or too much can dramatically alter the final outcome. Similarly, the precision in drying samples ensures they're in the perfect state for further preparation and eventual observation.

For many, the initial foray into mounting samples on slides is a rite of passage. Whether you're dealing with a solid or liquid specimen, learning to place it correctly on the slide is a skill that sets the stage for all the exploration to come. It's like learning the proper grip on a paintbrush; it might not make you a master painter immediately, but it's an essential step on the journey to creating beautiful art.

As you move through these initial steps, you'll find that each one is a building block, designed not only to prepare your sample for observation but also to prepare you for the intricate world of microscopy. With patience and practice, these basic sample preparation techniques will become second nature, allowing you to focus on the wonders waiting to be discovered through your microscope lens. And remember, every great discovery in microscopy started with a well-prepared sample.

Basic Sample Preparation Techniques

- Keeping It Simple

Diving into the world of microscopy can seem daunting at first, but mastering the basic sample preparation techniques can simplify the journey, making the microscopic world more accessible to everyone. These foundational steps are your toolkit for embarking on this exciting exploration, empowering you with the confidence to prepare your specimens for the microscopic stage.

The process begins with a meticulous cleaning of your samples. This might involve gentle washing or brushing to ensure that every speck of dust or unwanted particle is removed. Imagine you're an archaeologist, carefully uncovering hidden treasures; in microscopy, your treasure is the true nature of your sample, unveiled in its purest form.

Once your specimen is clean, the next step may involve drying. However, this is not always required and depends greatly on the nature of your sample and the type of microscopy you'll be engaging with. For those specimens that do need drying, the method chosen should be one that preserves their integrity. Think of this as preparing a canvas for painting; the surface needs to be just right for the masterpiece to take shape.

After your sample is prepped and ready, it's time to move on to mounting. This step is where you'll learn the finesse of placing your specimen on a slide in such a way that it can be viewed and analyzed under the microscope. Whether it involves using a drop of mounting medium or carefully positioning the specimen with tweezers, this technique requires a steady hand and a keen eye for detail. It's akin to setting a gemstone; the beauty is in the precision of its placement.

For those starting their microscopy journey, practicing these basic sample preparation techniques is like learning the notes before playing a symphony. It lays the groundwork for all the fascinating discoveries to come, turning the complex into the manageable.

As you become more comfortable with these steps, you'll find yourself more equipped to tackle the challenges of more advanced techniques, all while maintaining a sense of wonder and curiosity about the microscopic world that surrounds us. So, grab your tools and let's simplify the intricate process of preparing your samples for the magical journey under the microscope.

Staining Methods - Adding Color to Your Microscopic World

Imagine venturing into a vast, uncharted forest with the mission of mapping its diverse flora and fauna. Now, think of staining methods in microscopy as your set of vibrant markers, each color chosen to highlight different aspects of the microscopic landscape before you. These techniques are not just about adding visual appeal; they serve as crucial navigational tools that bring clarity to the complex and often hidden world under the microscope.

In the realm of microscopy, staining breathes life into otherwise transparent or faintly visible structures, allowing us to see the unseen. Each stain has its unique affinity for specific components within a sample, acting like a spotlight that illuminates the features we wish to examine closely. For instance, Gram staining, a technique used in microbiology, categorizes bacteria into two groups based on their cell wall properties, making it easier to identify and study them.

Embarking on this colorful journey requires a thoughtful selection of stains, each chosen for its ability to reveal specific details. Just like an artist selects their palette to convey a certain emotion or highlight a particular aspect of their subject, a microscopist chooses their stains based on the information they seek. Some stains excel in showcasing cellular structures, while others are perfect for highlighting nucleic acids or proteins.

The process of staining, however, is more than just applying colors. It demands precision, patience, and practice. Preparing your samples for staining involves a series of steps that must be carefully followed to ensure the stain adheres correctly and the underlying structure is preserved. It's a delicate balance between art and science, requiring a meticulous hand to achieve the desired outcome.

As you explore the application of various staining methods, you'll discover the incredible diversity of the microscopic world in vivid detail. From the deep blues of a methylene blue-stained slide revealing the intricate details of cell morphology, to the striking contrast of a Gram-stained sample uncovering the hidden world of bacteria, each stained slide is a window into the marvels of the microcosm.

Staining methods, therefore, are not just about adding color; they're about enhancing our understanding and appreciation of the microscopic universe. By mastering these techniques, you unlock the door to a world of discovery, where every color tells a story, and every stained slide holds the promise of new insights and revelations. So, let your curiosity guide you as you paint your way through the microscopic landscape, uncovering the beauty and complexity that lies within.

Mounting Samples - The Art of Presentation

Mounting samples in microscopy isn't just a necessary step; it's an art form that turns your slide into a stage, showcasing your specimens to their utmost potential. This technique is the final touch that brings your meticulous preparation full circle, ensuring that the specimen is not only visible under the microscope but presented in a way that highlights its features beautifully and accurately.

Think of it as framing a photograph. Just as the right frame can enhance the beauty of a picture, proper mounting accentuates the details of your specimen, making them clearer and more vivid under the microscope's lens. It's about creating the perfect environment for your sample to shine, providing a clear window into the microscopic world.

The process of mounting samples requires a careful and thoughtful approach. Each specimen, with its unique characteristics, may require a specific type of mounting medium – be it aqueous for water-soluble samples or non-aqueous for those that are not. It's akin to choosing the right setting for a gemstone, ensuring that its beauty is displayed to its best advantage.

Mounting is not just about aesthetics, though. It plays a crucial role in preserving the sample's integrity during the observation process. By securing the specimen on the slide, you prevent any movement that could disrupt your view or damage the delicate structures you're trying to observe. It's a protective embrace, safeguarding your specimen against the pressures of examination.

In addition to choosing the right medium, the act of mounting itself demands precision. Whether you're using a drop of mounting medium to envelop your sample in a protective bubble or carefully placing a cover slip to avoid air bubbles and ensure even distribution, your movements must be both deliberate and gentle. This step is where patience becomes your greatest ally, as rushing can lead to mistakes that obscure the very details you're striving to reveal.

Mastering the art of mounting samples is a testament to your dedication to microscopy. It reflects a commitment to not only observing the microscopic world but presenting it in the most illuminative and respectful way possible. As you refine your techniques in mounting, you'll find that this step is not just about preparation but about honoring the incredible complexity and beauty of the specimens you explore.

Preparing Wet Mounts and Dry Mounts - A Tale of Two Techniques

Embarking on the journey of microscopy invites a fascinating decision at the crossroads of sample preparation: choosing between the paths of wet mounts and dry mounts. This choice, far from being arbitrary, hinges on the nature of your specimen and the story you hope to unveil under the microscope. Each technique, with its own set of advantages and applications, illuminates a different aspect of the microscopic world, offering a window into the unseen with clarity and precision.

Wet mounts, akin to setting a scene under water, involve immersing your specimen in a liquid medium, such as water or glycerin, and then gently covering it with a cover slip. This technique breathes life into living specimens, allowing you to observe them in a state closest to their natural environment. It's particularly captivating when studying aquatic samples or observing the mobility of microorganisms, as the liquid medium provides the necessary conditions for life to flourish under your gaze. The enchantment of watching these microscopic beings move and interact in real-time is akin to peering into a living, breathing miniature world.

On the other hand, dry mounts offer a different perspective, one that's straightforward yet equally mesmerizing. Here, the specimen is placed directly onto the slide without any liquid. This method is especially suited for samples that are already dry or for those where moisture could obscure the details you seek to explore. Imagine placing a delicate feather, a pollen grain, or a thin slice of cork on the slide; each stands as a testament to the beauty and complexity of dry specimens. The absence of liquid ensures that structures remain unaltered, presenting a clear, unobstructed view of the specimen's architecture.

Both techniques, with their unique methods of showcasing specimens, invite you to adapt your approach based on the story you wish to tell through your microscope. Whether you choose the dynamic, life-sustaining environment of a wet mount or the stark, unembellished clarity of a dry mount, each method opens up a realm of discovery. As you navigate through the preparation of wet and dry mounts, you'll learn to appreciate the versatility and adaptability required in microscopy, embracing each technique as a tool to reveal the hidden wonders of the microscopic world.

Advanced Preparation Techniques - Taking It a Step Further

As your curiosity and passion for microscopy deepen, venturing into advanced preparation techniques becomes an exhilarating next step. Embracing these sophisticated methods not only expands your toolkit but also enriches your understanding of the microscopic universe. Through processes like sectioning, fixation, and embedding, you're granted access to a more profound and detailed exploration of your samples, unveiling aspects that remain hidden to more basic techniques. It's akin to acquiring a magnifying glass that reveals finer details of a beautiful tapestry, each thread and color more vivid and meaningful.

Embarking on sectioning, you learn the delicate art of slicing your specimens into thin, translucent pieces, perfect for microscopic examination. This method is much like carefully peeling away layers to reveal the core of a story. Each slice presents a new narrative about the internal structures and intricate designs of your specimens, allowing for an intimate glimpse into their inner workings. Sectioning becomes a dance of precision and patience, where the reward is a deeper insight into the microscopic architecture that composes our world.

Meanwhile, the practices of fixation and embedding serve as the guardians of your specimens 'integrity. Fixation is a meticulous process where samples are treated with specific chemicals to preserve their cellular and tissue structures. It's like capturing a moment in time, ensuring that the dynamic beauty of your specimens is preserved for eternity. Following fixation, embedding offers a supportive embrace to your samples, encasing them in a stable medium that facilitates easier handling and cutting for microscopic examination. Together, these processes safeguard the essence of your specimens, allowing you to explore their complexities without compromise.

Advanced preparation techniques, therefore, are not just steps in a process but a journey into the heart of microscopy. They demand a blend of precision, understanding, and respect for the microscopic world, challenging you to push the boundaries of what can be seen and understood. As you master these techniques, you become not just an observer but a storyteller, bringing to light the unseen stories woven into the fabric of the natural world.

Sectioning - A Closer Look Into the Structure

Sectioning opens the door to a hidden universe tucked away within the confines of our samples. It's an art that requires not just a steady hand, but an eye for the detail and beauty that lies beneath the surface. By slicing our specimens into thin, almost translucent layers, we're able to peel back the layers of the microscopic world, revealing its complex inner architecture with unparalleled clarity.

Think of sectioning as embarking on a treasure hunt, where each slice brings you closer to uncovering the secrets hidden within your samples. It's about gently carving through the specimen to expose the intricate details and patterns that tell the story of its structure and function. This process isn't just cutting for the sake of cutting; it's a methodical journey into the essence of the specimen, guided by curiosity and the quest for knowledge.

The true beauty of sectioning lies in its ability to bring to light the delicate dance of cells and tissues, structures too intertwined and complex to be fully appreciated without this finer examination. Just as a sculptor removes layers of stone to reveal the form within, sectioning allows us to strip away the outer layers of our specimens, bringing their inner beauty into focus. It's a technique that transforms the invisible into the visible, the overlooked into the centerpiece of our microscopic exploration.

With each slice, we're not just cutting through material; we're uncovering layers of meaning, diving deeper into the life of the specimen. This process enables us to traverse the microscopic landscape, exploring valleys and mountains of cellular structures that form the fabric of life itself. Sectioning isn't just about observing; it's about connecting on a deeper level with the microscopic world, understanding its complexities, and appreciating the beauty that lies within the minutiae.

In the realm of microscopy, sectioning serves as a bridge between the seen and the unseen, providing a pathway to discoveries that enrich our understanding of the microscopic universe. It's a testament to the wonders that await us when we take the time to look closer, to explore further, and to appreciate the profound intricacies of the world at the microscopic level.

Fixation and Embedding - Preserving Your Samples for Posterity

Navigating the journey of microscopy, we arrive at a pivotal milestone: the twin practices of fixation and embedding. These techniques are not just procedural steps; they're acts of preservation, safeguarding the delicate architecture of your samples for the long haul. Fixation acts as the guardian of cellular structure, halting the processes that could lead to degradation. Imagine capturing a fleeting moment in perfect clarity—a sunset paused forever in its breathtaking beauty. Similarly, fixation freezes your specimens in their moment of perfection, ensuring their intricate details are maintained for future exploration.Following fixation, embedding provides a supportive haven for your specimens. It's akin to placing a precious artifact within a protective case, ensuring it's shielded from harm while still allowing for detailed examination. This process surrounds your sample with a stable matrix, often paraffin or resin, offering the support needed for thin sectioning without compromising the specimen's integrity. Think of it as crafting the perfect setting for a gemstone, enhancing its beauty and allowing its details to shine under scrutiny.Together, fixation and embedding form a duo of preservation, transforming your specimens into timeless pieces ready for the microscopic stage. They allow you to delve into the past with the tools of the present, exploring the preserved beauty of biological structures as if they were still living. By mastering these techniques, you ensure that the wonders of the microscopic world are not just a fleeting glimpse but a lasting legacy, ready to be revisited and appreciated time and again. Through the art of preserving your samples, you bridge the gap between the now and the future, crafting a foundation for discoveries that transcend time.

CHAPTER 6: MASTERING YOUR MICROSCOPE, TIPS, TRICKS AND TROUBLESHOOTING

Using a Microscope

Embarking on your microscopy journey can initially feel like navigating a new universe. However, understanding the basics of using a microscope can transform this complexity into an exciting adventure. Microscopes are magnificent tools that extend our sight into the miniature worlds around us, revealing details and secrets invisible to our eyes alone. This section is your first step toward becoming adept at maneuvering through the microscopic realm, focusing on practical insights to enhance your skills.

When you approach a microscope, think of it as meeting a new friend. Familiarize yourself with its parts: the eyepiece, where you'll be looking into the hidden micro-world; the objective lenses, which are the primary tools for magnification; the stage, where you'll place your samples; and the focus knobs, which you'll adjust to bring your samples into sharp clarity. Recognizing these components and their roles is the foundation of your microscopy mastery.

Adjusting to the proper use of a microscope is akin to learning a new language—it might seem challenging at first, but with practice, it becomes second nature. Begin by practicing how to correctly place a slide on the stage and secure it with the stage clips. This simple action is crucial, as it prevents the slide from moving while you're exploring its wonders.

As you proceed, remember the importance of starting with the lowest magnification. This approach allows you to locate your sample more easily and gradually zoom in for finer details. It's an essential practice that protects both your specimen and the microscope's lenses from potential damage.

Another pivotal skill is mastering the art of focusing. It might be tempting to quickly twist the focus knobs, but patience and gentle adjustments are your allies here. Start with the coarse focus to bring your sample into a general view, then use the fine focus for precision and clarity. This methodical approach not only ensures the best possible image but also helps to preserve the integrity of your equipment.

Exploring the microscopic world requires a blend of curiosity, patience, and respect for the instrument in your hands. Each sample presents a unique opportunity to discover something new, making every viewing experience a valuable part of your learning journey. As you spend more time with your microscope, you'll develop a keen eye for detail and a deeper appreciation for the intricate beauty of microscopic structures.

Remember, proficiency in microscopy comes with practice and patience. Embrace each step of the process, and don't hesitate to experiment within the bounds of proper care and handling. With every slide you examine, you're not just observing—you're connecting with a fascinating world beyond the reach of the naked eye. Keep this adventurous spirit alive as you continue to explore, learn, and grow in your microscopy journey.

Setting Up Your Microscope

Embarking on the adventure of microscopy begins with the crucial step of setting up your microscope properly. This initial setup is not just about assembling parts—it's about creating a foundation for all the incredible discoveries to come. Imagine yourself as the captain of a ship, preparing for a voyage into the unknown. Just as a ship must be carefully prepared before setting sail, so too must your microscope be meticulously set up to ensure a successful journey into the microscopic world.

First, find a stable and clean surface to place your microscope. This space should be away from any potential disturbances that could shake or shift the microscope during observation. Think of this space as your command center, a place where focus and precision are paramount. Once you have identified the perfect spot, it's time to get acquainted with your microscope's components. You'll want to assemble any parts that aren't already in place, such as attaching the eyepiece or objective lenses if they're not already secured.

Next, connect the microscope to a power source if it requires one, and then switch on the light source. The light is the beacon that will illuminate your path through the microscopic landscape, so adjusting it correctly is essential. Begin with a low light setting and adjust as necessary depending on the transparency and nature of your samples.

After the light source is set, position the microscope so that the stage is ready to receive slides. If your microscope has a mechanical stage, familiarize yourself with the controls. These will allow you to move your sample smoothly in both the X (left and right) and Y (forward and backward) directions. This smooth navigation is crucial for scanning your specimen and finding areas of interest.

With the stage prepared, it's time to select your starting objective lens. Always begin with the lowest magnification. This makes it easier to find your sample when you place it on the stage. Carefully clip your slide onto the stage, ensuring it's secure but not forcibly pressing down on it. Remember, the slide is your window into the microscopic world, and treating it with care is paramount.

Once your slide is in place, use the coarse adjustment knob to get the initial focus. Look through the eyepiece as you slowly adjust, watching as shapes and colors start to emerge from the blur. This moment, when the hidden world first comes into view, is nothing short of magical.

As you proceed with these steps, you're not just setting up a microscope—you're setting the stage for discovery, for the thrill of exploration, and for a deeper understanding of the world at a scale that's usually out of sight. Each step in the setup process is an essential part of your journey into microscopy, paving the way for countless hours of learning and exploration.

Proper Handling and Care

In the journey of mastering microscopy, proper handling and care of your microscope stand as fundamental practices that ensure the longevity and reliability of this sophisticated instrument. Embrace the responsibility of caring for your microscope with the same enthusiasm you have for exploring the microscopic world. It's not just about keeping the device clean; it's about preserving a gateway to infinite discoveries.

Begin by always keeping your microscope covered when not in use. Dust and environmental particles can be the greatest adversaries of optical clarity, potentially clouding lenses and complicating your view of the microcosm. Utilize a dedicated, soft, lint-free cloth or brush for gently removing any dust or debris from the lenses and body of the microscope. Avoid using regular household cleaning agents; instead, opt for solutions specifically designed for optical components to prevent damage.

Handling your microscope with care also extends to its transportation and storage. When moving the microscope, do so by gripping the arm with one hand while supporting the base with the other. This ensures stability and reduces the risk of accidental drops or bumps. Find a safe, dry, and cool storage area away from direct sunlight and excessive humidity, as these conditions can deteriorate the microscope's mechanical and optical parts over time.

Maintenance is another key aspect of proper microscope care. Regularly check and tighten any loose screws or moving parts to keep the microscope functioning smoothly. Be vigilant about the condition of the electrical components if your microscope has a built-in light source. Ensure that cords are intact and free from frays, and always disconnect the power supply before conducting any maintenance or cleaning.

When it comes to cleaning the optical elements, such as eyepieces and objective lenses, exercise extra caution. Use only special optical cleaning solution and lens paper to wipe them gently. Any scratch or residue on these components can significantly impair image quality, detracting from the very essence of microscopy.

Moreover, fostering a habit of inspecting your microscope before and after use can catch potential issues early, preventing them from escalating into bigger problems. This includes checking for any loose components, ensuring smooth operation of focus knobs, and verifying that the light source functions correctly.

Embracing these practices of proper handling and care not only augments your microscopy experience but also embodies a respect for the instrument that unlocks the door to a microscopic universe. Treat your microscope as a valued companion in your exploratory journey, and it will serve you faithfully, unveiling the marvels of the unseen world with clarity and precision.

Adjusting the Light Source and Focus

Illuminating the microscopic world in just the right way can make all the difference in what you're able to see and discover. Properly adjusting the light source and focus on your microscope might seem like a simple task, but it's a critical skill that can elevate your microscopy experience from good to great. Here's how to get it just right, shedding light on the wonders of the microscopic universe with clarity and precision.

First off, let's talk about your microscope's illumination. If your microscope uses a built-in light source, start by turning it on at its lowest setting. This initial step helps to protect both your eyes from sudden brightness and your specimen from potential heat damage. Gradually increase the light intensity until you achieve a level that brightly illuminates your sample without washing out the delicate details you're aiming to observe. If your microscope is equipped with a condenser and diaphragm, adjusting these can further refine the quality of light that reaches your specimen, enhancing contrast and detail in your view.

Now, onto focusing, a pivotal step that brings the hidden details of your samples into sharp relief. Begin with the lowest magnification objective lens in place. This not only gives you a broader view of your specimen but also minimizes the risk of damaging your slide or lens. Place your eye close to the eyepiece and slowly rotate the coarse focus knob until the outlines of structures start to emerge from the blur. This is your cue to switch to the fine focus knob for the detailed work. With a gentle touch, dial in the fine focus until you achieve a crisp, clear image.

The dance between adjusting light and focus is a delicate one. Too much light can obscure fine details, while too little can leave you straining to see. Similarly, hastily adjusting the focus can cause you to miss the subtleties that make each sample unique. Patience and gradual adjustments are your best tools here.

Experiment with different levels of illumination and focus adjustments to find the perfect balance for each specimen you examine. Remember, what works best for one sample may not be ideal for another. Through practice, you'll develop an intuitive sense for how to adjust the light source and focus to reveal the incredible detail and beauty of the microscopic world. Embrace the process of experimentation and discovery, and you'll find that adjusting the light source and focus becomes a natural extension of your curiosity and passion for microscopy.

Viewing Samples

Stepping into the phase of viewing samples through your microscope marks the moment where preparation meets discovery. This exciting stage is where your skills in setup and adjustment converge, allowing you to delve into the intricacies of the microscopic world. However, knowing how to adeptly place and prepare your specimens is crucial for embarking on this investigative journey.

To begin, select a clean slide and, if necessary, a coverslip. Your sample should be thin and evenly spread if it's a solid or liquid specimen. This ensures that the light can easily pass through, providing a clear view under the microscope. For solid objects, a thin section sliced with precision will offer the best results. Meanwhile, liquid samples often require a drop placed in the center of the slide, gently covered with a coverslip to avoid air bubbles, as these can obscure your view.

Mounting your specimen properly is the next critical step. Position your slide on the stage and secure it with the stage clips. This steadiness is key to preventing any unwanted movement that could blur your vision during observation. Once secured, start with the lowest magnification objective lens. This broader perspective makes locating your specimen easier before zooming in for a closer look.

As you explore, remember to adjust the light and focus gradually, as detailed in previous sections. Each sample might require a different level of illumination or focus adjustment, so flexibility and patience are vital. Experimenting with these settings can unveil hidden details or highlight features of your specimen that are not immediately obvious.

When moving to higher magnifications, re-center your specimen if it drifts out of view. This often happens when switching objective lenses due to the varying fields of view. It's a normal part of the process and a good opportunity to refine your focusing skills further.

Exploring the microscopic realm is akin to venturing into an art gallery, where every specimen displays its unique patterns and structures, inviting you to observe and interpret. Some samples might reveal their secrets readily, while others may require time and adjustments to uncover their full story. Approach each viewing experience with curiosity and an open mind, ready to adjust and explore.

As you become more comfortable with viewing samples, you'll find that this process not only enhances your understanding of microscopy but also deepens your appreciation for the minute wonders that make up our world. Each slide offers a new story to tell, a mystery to solve, or a phenomenon to marvel at, making every viewing experience a pivotal moment in your microscopy adventure.

Tips for Beginners

Embarking on the microscopy journey can seem daunting at first glance, but with the right guidance and a few insider tips, beginners can swiftly move from novices to adept users, unlocking the microscopic world's vast mysteries. Here are practical suggestions designed to smooth your path and enrich your microscopy experience.

First, prioritize learning the basic anatomy of your microscope. Understanding each part's function and location not only aids in proper usage but also in troubleshooting should issues arise. Familiarize yourself with terms like 'eyepiece,' 'objective lenses,' 'stage,' and 'focus knobs'—this foundational knowledge is crucial for effective microscopy.

It's also essential to cultivate patience. Patience is your greatest ally when adjusting the focus or changing the magnification. Rushing through these adjustments can lead to frustration and may even damage your samples or the microscope itself. Allow yourself the time to slowly and carefully make adjustments, and with practice, these tasks will become more intuitive.

Practicing good slide preparation is another key skill. Begin with simple specimens and strive for thin, evenly spread samples. This will not only make viewing easier but also help in understanding how different specimens react under magnification and how adjustments to lighting and focus can reveal intricate details.

Another valuable tip is to document your observations. Keeping a journal of what you view, including sketches or descriptions of the samples, the magnification used, and any peculiarities noticed, can be incredibly beneficial. This practice not only enhances your observation skills but also creates a personal reference that you can look back on to track your progress and insights.

Ensure that you're also taking breaks, especially during prolonged viewing sessions. Extended periods of peering through a microscope can lead to eye strain. Periodic breaks help maintain your focus and enthusiasm, ensuring that each microscopy session remains a pleasure rather than a chore.

Lastly, embrace the community. Whether online forums, social media groups, or local clubs, connecting with other microscopy enthusiasts can provide support, inspiration, and answers to your questions. The collective wisdom of a community can be a rich resource for learning and discovery.

By following these tips and approaching your microscopy journey with curiosity and openness to learning, you'll find that what initially seemed complex becomes an engaging and rewarding pursuit. Remember, every expert was once a beginner—your microscopy adventure is just beginning, and there's a whole microscopic universe waiting to be discovered.

Common Problems and Troubleshooting

Navigating the occasional hiccup with your microscope is an inevitable part of the learning curve. But don't let these setbacks dampen your spirit; instead, view them as opportunities to deepen your understanding of your instrument. Here we'll explore some frequently encountered issues and offer solutions to keep your microscopy journey on track.**Problem: Blurry or Unclear Images**This is perhaps the most common challenge faced by microscope users. Start by checking if the objective lens and eyepiece are clean. A simple cleaning with lens paper might resolve the issue. Next, ensure that your sample is properly prepared and positioned. If the problem persists, adjusting the light source or fine-tuning the focus can often bring the clarity you're seeking.**Problem: Difficulty in Locating Specimens**Especially at higher magnifications, finding your specimen can feel like searching for a needle in a haystack. Always begin your observations under the lowest magnification to locate the specimen more easily and then gradually increase magnification.**Problem: The Light Source Isn't Working**First, verify that your microscope is properly plugged in and the power switch is on. If you're using a battery-operated model, check if the batteries need replacing. For microscopes with a built-in light source, ensure that the bulb is not burnt out and securely fastened.**Problem: Mechanical Parts Seem Stuck**If the focus knobs or stage adjusters are not moving smoothly, refrain from forcing them. This could be due to over-tightening or debris accumulation. Gently clean visible debris and, if needed, consult the manufacturer's guide on lubrication procedures suitable for your microscope model.**Problem: Eye Strain from Extended Use**Microscopy is as much about technique as it is about endurance. Remember to take frequent breaks during long sessions to prevent eye strain. Adjusting the eyepiece diopter, if available, to match your vision can also provide a more comfortable viewing experience.Each of these challenges, while potentially frustrating, serves as a stepping stone toward becoming proficient in microscopy. With a bit of troubleshooting, you can overcome these hurdles and continue your exploration of the microscopic world with confidence and curiosity.

CHAPTER 7: DIVING DEEP INTO MICROSCOPY, EXPLORING SPECIALIZED MICROSCOPES

The Magical World of Fluorescence Microscopes

Embark on an enchanting journey into the magical world of fluorescence microscopy, a realm where light transforms the microscopic landscape into a vivid tableau of life. At the heart of this technique lies the art of using fluorescence to shed light on the intricate details of cells and tissues, allowing us to visualize the building blocks of life like never before.

Imagine a world where we can highlight specific components within a cell, thanks to the marvel of fluorescent dyes and proteins. These markers glow under the microscope, revealing the dynamic dance of life at a molecular level. Each color tells a story, mapping the whereabouts of DNA, proteins, and other vital molecules, thus providing us with a detailed map of cellular function and structure. This ability to pinpoint precise elements within complex biological systems is what sets fluorescence microscopy apart, turning it into a cornerstone of modern biological research.

The applications of fluorescence microscopy are as diverse as they are groundbreaking. In the realm of medical research, it plays a pivotal role in understanding diseases, from unraveling the mysteries of cancer cell behavior to observing the intricate interplay of viruses and host cells. This technique also shines in developmental biology, where it illuminates the stages of embryonic development with unprecedented clarity, allowing scientists to witness the miracle of life as it unfolds.

But the magic of fluorescence microscopy doesn't stop at observation. It's a gateway to experimentation, enabling researchers to track the effects of drugs within cells, understand gene expression patterns, and explore the mechanics of cell signaling pathways. This dynamic tool not only allows us to see the unseen but also to ask questions and seek answers about the fundamental processes of life.

Yet, the true enchantment of fluorescence microscopy lies in its constant evolution. With each advancement in technology, from improved fluorescent markers to advanced imaging systems, we peel back another layer of the microscopic world. Innovations such as time-lapse imaging bring the cellular world to life, offering a glimpse into the bustling activity within cells over time. These technological leaps forward expand our understanding and push the boundaries of what's possible in scientific research.

Fluorescence microscopy, with its blend of art and science, invites us into a world where the smallest details matter. It's a world where the unseen becomes visible, and the complexities of life are brought into the light. As we continue to explore and innovate, the magical world of fluorescence microscopy promises to be at the forefront of discovery, opening new doors to knowledge and inspiring a sense of wonder in the beauty of the microscopic world.

Going Deeper with Confocal Microscopy

Imagine being able to peel back the layers of a specimen, diving into its very essence to reveal secrets hidden from the naked eye. This is the power of confocal microscopy, a technique that transforms the way we observe the microscopic world. By utilizing a specialized pinhole aperture, confocal microscopy sharpens our view, slicing through the blur to bring into focus the vibrant, intricate details of life at the cellular level.

This innovative approach to microscopy allows for the exclusion of out-of-focus light, which is a common obstacle in traditional fluorescence microscopy when examining thicker specimens. The result? Crisp, high-resolution images that are rich in contrast and detail, enabling researchers to observe structures and processes within cells and tissues with an unprecedented clarity. It's like having the ability to focus precisely on each individual layer of a complex, multi-layered cake, appreciating every detail and texture without the interference of the surrounding layers.

One of the most captivating applications of confocal microscopy is its ability to construct three-dimensional images of biological samples. By compiling multiple, thin, optical sections along the z-axis, scientists can assemble a comprehensive 3D reconstruction of their specimen. This capability opens a window into the architectural beauty of biological structures, from the intricate network of neurons in the brain to the delicate framework of blood vessels within an organ. Researchers can virtually navigate through these 3D landscapes, gaining insights into the spatial relationships and interactions that govern life at the microscopic scale.

Confocal microscopy has carved its niche across a spectrum of scientific endeavors. In neuroscience, it illuminates the complex wiring and synaptic connections of the brain, offering clues to understanding neural networks and brain functions. Cell biologists rely on it to dissect the inner workings of cells, studying phenomena such as organelle dynamics, protein trafficking, and cell signaling. Developmental biologists, too, find it invaluable for tracking the stages of embryonic development, witnessing the emergence of life's complexity from a single cell.

Beyond the realm of biology, confocal microscopy extends its reach into materials science, where it aids in the characterization of polymers, fibers, and other composite materials. Its ability to provide detailed surface topology and composition analysis makes it a tool of choice for innovators designing the next generation of materials.

However, the journey into the depths of the microscopic world with confocal microscopy is not without its challenges. Mastering this technique requires skill and patience, as well as an understanding of its complexities. Researchers must navigate the delicate balance between resolution, signal-to-noise ratio, and sample integrity. Each decision, from selecting the appropriate laser wavelength to adjusting the pinhole size, must be carefully considered to optimize image quality without compromising the specimen.

Despite these challenges, the rewards of confocal microscopy are immense. It provides a gateway to exploring the unknown, pushing forward the frontiers of science and medicine. As we continue to refine and innovate upon this technique, its potential to unravel the mysteries of life at the microscopic level grows ever more promising. Confocal microscopy not only enhances our capacity to see but to understand, offering a deeper connection to the minuscule wonders that make up our world.

The Precision of Atomic Force Microscopy (AFM)

Step into the world of Atomic Force Microscopy (AFM), a technique that veers away from traditional microscopy and takes us on an intimate journey across surfaces at the nanoscale. This realm of microscopy is less about observing through a lens and more about feeling the contours, textures, and forces at play on a surface. With a fine probe, akin to a microscopic finger, the AFM gently scans across a sample, mapping out its topography with astounding detail and precision.

This method of microscopy opens up a unique window to a world unseen by optical methods, offering insights into the structure and properties of materials at an atomic level. It's akin to reading Braille, where the shape and size of each letter are discerned by touch, allowing us to "see" the nanoworld through touch. The probe of an AFM can detect variations in height as minute as a fraction of a nanometer, bringing the bumps and grooves of surfaces into sharp relief.

Beyond merely mapping surfaces, AFM possesses the extraordinary ability to measure the forces between the probe and the sample. This includes detecting the subtle push and pull of atomic and molecular interactions, providing a detailed picture of the mechanical properties of materials. These capabilities make AFM an invaluable tool in fields as diverse as materials science, where it aids in the design and testing of new materials, and biology, where it allows the examination of cells, proteins, and DNA in their native environments.

The application of AFM extends beyond static imaging. It enables the study of dynamic processes, such as how a material's structure changes under stress or in different environmental conditions. Researchers can observe in real-time how individual atoms and molecules on a surface respond to external forces, offering insights into phenomena at the most fundamental levels of physics and chemistry.

Furthermore, AFM is not limited to conductive materials; it can image biological specimens and soft materials without the need for special treatments that could potentially alter their native states. This makes it a powerful tool for biophysicists and bioengineers, who can explore the mechanics of cell membranes, the assembly of protein complexes, and the behavior of DNA strands with an intimacy and detail that were previously unattainable.

Despite its impressive capabilities, navigating the world of AFM requires a steady hand and a nuanced understanding of the technique. The interpretation of AFM data can be complex, as the images are the result of mechanical interactions at the nanoscale. Careful calibration and control of the probe, as well as a deep understanding of the sample's properties, are essential to ensuring the accuracy and relevance of the findings.

AFM represents a convergence of physics, engineering, and nanotechnology, allowing us to explore the building blocks of materials and life with unparalleled clarity. While challenges in technique and interpretation exist, the potential of AFM to uncover the secrets of the nanoscale world is boundless. As we continue to refine this method and expand its applications, the journey into the depths of surfaces at the atomic level promises to reveal new mysteries and opportunities, pushing the frontier of our understanding ever forward.

Applications of Specialized Microscopes in Research

The realm of specialized microscopy has unlocked a treasure trove of possibilities across various scientific disciplines, each microscope type bringing its unique lens to the table. These tools have not only enabled us to venture beyond the limits of traditional microscopy but have also catalyzed significant breakthroughs in both our understanding and innovation within numerous research fields.

In the vibrant field of cellular biology, fluorescence microscopes have become indispensable. By harnessing the power of fluorescent markers, researchers can illuminate the dance of proteins, track the flow of genetic information, and observe the dynamic interactions within cells in real time. This capability to tag and visualize specific molecules within cells has shed light on the molecular underpinnings of countless biological processes and diseases, offering clues to potential therapeutic targets.

Delving into the architectural complexities of biological tissues, confocal microscopes have emerged as a critical tool. With their ability to peer into the depths of specimens and construct detailed three-dimensional images, confocal microscopy has provided invaluable insights into the structural organization of tissues and organs. This has profound implications for understanding developmental biology, neuroanatomy, and the progression of various diseases, such as cancer, where the microenvironment's spatial configuration plays a crucial role.

Atomic force microscopy (AFM), on the other hand, has revolutionized our approach to nanoscale exploration. By feeling the surface of samples at the atomic level, AFM has opened new avenues for material science, offering a detailed understanding of surface properties, mechanical characteristics, and molecular interactions. This has not only advanced the development of novel materials with tailored properties but also provided a deeper understanding of biological membranes and protein folding mechanisms, bridging the gap between physical sciences and biology.

Beyond these individual contributions, specialized microscopes have fostered interdisciplinary research, bringing together chemists, physicists, biologists, and engineers. For instance, the convergence of fluorescence and atomic force microscopy has enabled researchers to correlate mechanical properties with biochemical behavior within cells, offering a more holistic view of cellular mechanics. Similarly, the integration of confocal microscopy with other imaging modalities is enhancing our understanding of complex biological systems, allowing for a more nuanced exploration of disease mechanisms and potential therapeutic interventions.

The applications of specialized microscopes extend into the realm of diagnostics and disease research, where they play a pivotal role in identifying pathogens, understanding disease progression, and developing new treatments. For example, fluorescence microscopy has been crucial in studying the interactions between host cells and pathogens, shedding light on the mechanisms of infection and immune response. Meanwhile, confocal microscopy's ability to generate high-resolution images of tissue sections is invaluable for histopathology, helping diagnose and understand diseases at a cellular level.

In essence, the diverse applications of specialized microscopes have not only expanded the frontiers of scientific research but have also paved the way for new discoveries and innovations. By providing a window into the unseen world, these tools continue to enrich our knowledge, challenge our assumptions, and inspire curiosity, driving the pursuit of discovery across the vast landscape of science.

Challenges and Limitations

Embarking on the journey of specialized microscopy invites us to confront a unique set of challenges and limitations, each demanding our awareness and innovative spirit. As we delve into the microscopic world with tools like fluorescence, confocal, and atomic force microscopes, we encounter hurdles that remind us of the complexity and delicacy of our quest for knowledge.

In the vivid arena of fluorescence microscopy, one significant challenge is the photobleaching of fluorophores. This phenomenon, where fluorescent molecules lose their ability to emit light upon prolonged exposure, can hinder long-term observations and diminish the quality of our data. Additionally, distinguishing our fluorescent signals from background noise requires meticulous technique and careful preparation of samples to ensure clarity and precision in our results.

Turning our gaze to confocal microscopy, its strengths in

providing detailed images are balanced by considerations of time and expertise. The meticulous process of acquiring sharp, focused images layer by layer can be time-intensive, demanding patience and precision. Moreover, mastering this technique necessitates a deep understanding of its principles and an adept handling of its sophisticated equipment, making it essential for users to undergo thorough training.

The realm of atomic force microscopy (AFM), with its ability to feel the contours of the nanoworld, introduces us to sensitivities to environmental conditions such as temperature and humidity. These factors can influence the behavior of both the probe and the sample, potentially affecting the accuracy of the measurements. The meticulous preparation of samples and the careful calibration of the instrument are crucial steps in overcoming these sensitivities to ensure the reliability of the data collected.

Beyond these specific challenges, all specialized microscopes share the common hurdle of data complexity. The richness of the information they provide comes with the task of interpretation, requiring a combination of technical skill, scientific knowledge, and sometimes, intuition to translate complex datasets into meaningful insights. This complexity often necessitates collaboration across disciplines, bringing together experts in microscopy, biology, physics, and computational analysis to fully harness the power of these advanced techniques.

Moreover, the rapid pace of technological advancement in microscopy presents both an opportunity and a challenge. Staying abreast of the latest developments and integrating new methodologies into existing research frameworks demand flexibility and a commitment to lifelong learning. Researchers and institutions must balance the excitement of innovation with the practicalities of training, funding, and infrastructure support to make the most of these evolving tools.

Despite these challenges and limitations, the pursuit of understanding through specialized microscopy remains a profoundly rewarding endeavor. Each obstacle overcome and each limitation navigated brings us closer to unraveling the mysteries of the microscopic world. With a blend of creativity, collaboration, and perseverance, researchers continue to push the boundaries of what is possible, turning challenges into stepping stones on the path to discovery. As we look to the future, our journey through the landscape of specialized microscopy is sure to uncover new horizons, enriched by the depth and diversity of our explorations.

The Future of Microscopy

The horizon of microscopy is expanding rapidly, driven by breakthroughs and innovations that promise to unlock even deeper secrets of the microscopic world. Imagine a realm where we're not just observing, but truly understanding the complexities of life at an unprecedented level. This is not just a dream; it's the future that's unfolding right before our eyes, thanks to the relentless progress in microscopy techniques.

One of the most thrilling advancements lies in the realm of super-resolution microscopy. This game-changing technology is shattering the previously accepted limits of optical microscopy, enabling us to visualize structures at the nanoscale that were once thought to be beyond our reach. By overcoming the diffraction limit, researchers can now delve into the intricate details of cellular structures with stunning clarity, opening up new avenues for biological discovery.

Equally transformative is the rise of single-molecule imaging. This powerful approach allows scientists to track the behavior and interactions of individual molecules within cells in real time. It's like having a front-row seat to the molecular dance of life, offering insights into the dynamics and mechanisms that underpin cellular function. This level of detail could revolutionize our understanding of complex biological processes and disease mechanisms.

The integration of artificial intelligence (AI) and machine learning

(ML) into microscopy is another frontier that's rapidly evolving. These technologies are not just enhancing the way we analyze microscopy data; they're transforming it. With AI and ML, we can sift through vast amounts of imaging data with unprecedented speed and accuracy, identifying patterns and insights that would be impossible for the human eye to detect. This not only accelerates research but also opens up new possibilities for personalized medicine and diagnostics.

Furthermore, the future of microscopy is not just about improving individual techniques but also about how they can be combined. The emergence of multi-modal imaging approaches is enabling researchers to merge different types of microscopy, each offering complementary views of the sample. This holistic approach provides a more comprehensive understanding of biological and material samples, from the macroscopic down to the molecular level.

As we peer into the future, the possibilities seem limitless. Yet, with these advancements comes the challenge of accessibility. Ensuring that these cutting-edge technologies are available to researchers around the world is crucial for driving global scientific progress. Efforts to democratize access to advanced microscopy will be key in shaping a future where the mysteries of the microscopic world are within reach of every curious mind.

In this journey toward the future, the microscopy community stands at the threshold of discovery, poised to explore the uncharted territories of the microscopic universe. With each technological leap, we're not just seeing further; we're connecting more deeply with the fabric of life itself. The future of microscopy promises not just to unveil the hidden wonders of the microcosm but to inspire a new generation of scientists and explorers, eager to uncover the secrets that lie waiting in the world beyond our sight.

CHAPTER 8: DECODING MICROSCOPY, A BEGINNER'S GUIDE TO CAPTURING AND ANALYZING IMAGES

Mastering Microscopy: Tips and Tricks for Capturing Perfect Images

Embarking on the journey of mastering microscopy can feel like unlocking a secret window into the unseen world. But, the key to truly unveiling this world lies in capturing those perfectly detailed images that tell a story of their own. Achieving such mastery isn't just about having the most advanced equipment; it involves a blend of technique, patience, and a bit of creativity.

Firstly, consider the cornerstone of microscopy: sample preparation. The way you prepare your slides can make a significant difference in the clarity and quality of your images. From ensuring specimens are thinly sliced for optimal light transmission to using the right staining methods to highlight crucial features, attention to these details sets the stage for successful imaging.

Lighting is another critical aspect that can dramatically affect the outcome of your photomicrography efforts. Whether you're working with transmitted light in brightfield microscopy or the glow of fluorescence, learning to adjust the intensity and direction of light can help in minimizing reflections and enhancing contrast. This might mean experimenting with different settings and observing how they influence the appearance of your specimen.

Focus is where the magic happens - or doesn't. The difference between a good image and a great one often boils down to how well-focused it is. Mastering the fine focus on your microscope requires a gentle touch and patience. Move slowly, observe the changes, and you'll find that sweet spot where every detail pops into sharp relief.

Choosing the right magnification is about finding balance. While it's tempting to zoom in as much as possible to catch every tiny detail, sometimes less is more. A lower magnification can provide a broader context of your specimen, making it easier to understand its structure and function. The goal is to select a level that reveals your subject's most telling details without losing perspective.

Lastly, embrace the learning curve. Every slide offers a new challenge and an opportunity to refine your skills. With each image captured, you'll learn a bit more about the intricate dance of light, focus, and magnification. Remember, the path to mastering microscopy is a journey, not a sprint. Enjoy the process, and before you know it, you'll be capturing those perfect images that unveil the mesmerizing beauty of the microscopic world.

Capturing Images Through a Microscope

Embarking on the fascinating journey of capturing images through a microscope is an exhilarating blend of science and art. It's like having the superpower to reveal the hidden intricacies of the microscopic world, but it does require a harmonious blend of knowledge, skill, and practice. Here's a friendly guide to help you navigate the nuances of this thrilling process.

First off, diving into the world of microscopic imaging isn't just about pointing and shooting; it involves understanding the unique language of light and shadows, shapes and colors. Each microscopy technique, whether it's the traditional brightfield, the contrast-enhancing phase contrast, or the vibrantly detailed fluorescence microscopy, opens up new avenues to explore and capture the unseen. Embracing these techniques means you're not just taking pictures; you're illuminating the hidden beauty of the microcosm.

A critical step in this journey is getting to know your equipment. Familiarizing yourself with your microscope, understanding its capabilities, and learning how to adjust its settings to suit your specimen and the chosen microscopy technique is vital. Each adjustment, from the light source to the objective lens, plays a pivotal role in the clarity and quality of your captured image. It's akin to tuning a musical instrument to get that perfect harmony.

The adventure continues as you delve into the world of camera setups. Whether you opt for a specialized microscope camera, a high-resolution DSLR, or even a smartphone adapter, the key is to choose a setup that aligns with your microscopy goals. It's about balancing the practical aspects, like resolution and sensitivity, with the creative ones, like framing and composition, to capture that perfect shot.

Practice is your best companion on this journey. It's through trial and error, experimenting with different specimens, lighting conditions, and magnifications, that you truly refine your skills. Each attempt is an opportunity to learn something new, to tweak a setting here, adjust a light there, all in the pursuit of that captivating image that tells a story, that sparks curiosity, that showcases the unseen world in all its glory.

So, take a deep breath, be patient, and let your curiosity guide you.

The path to capturing images through a microscope is filled with discoveries, challenges, and immense satisfaction. Remember, each image you capture is a testament to your dedication and a window into the vast, beautiful microscopic world waiting to be explored.

Photomicrography: An Art and Science

Photomicrography stands at the captivating crossroads where science meets art, offering a unique canvas that blends the precision of scientific inquiry with the boundless creativity of artistic expression. This fascinating discipline invites us to capture the unseen beauty of the microscopic world, transforming it into awe-inspiring imagery that can captivate both the scientific community and the public alike.

At its core, photomicrography requires not only a solid understanding of microscopy techniques but also an artistic eye capable of seeing beyond the ordinary. It's about recognizing the potential for beauty in the minute details of a cell structure or the intricate patterns found in crystalline formations. The journey into photomicrography begins with curiosity and a willingness to explore the possibilities that lie within a single drop of water, a piece of plant tissue, or a flake of mineral.

To excel in this art form, one must develop a deep appreciation for the interplay of light and shadow, color and contrast. Crafting an image that tells a compelling story involves making thoughtful decisions about composition, choosing the right magnification to highlight the most intriguing aspects of the subject, and manipulating light in ways that reveal hidden textures and dimensions.

The technical aspects of photomicrography, such as selecting the appropriate camera setup and mastering the nuances of imaging software, are critical to achieving high-quality results. However, the true essence of this art lies in the ability to envision the end result before the shutter clicks. It's about anticipating how the slightest adjustment in lighting or focus can dramatically alter the mood and impact of the image.

Experimentation is the heartbeat of photomicrography. Embracing trial and error, pushing the boundaries of traditional microscopy, and continually seeking new ways to showcase the microscopic world are what fuel progress in this field. Each successful image serves not only as a testament to the photographer's skill and vision but also as an invitation for others to look closer, to question, and to marvel at the wonders of the world on a scale too small for the eye to see.

Photomicrography, in essence, is a celebration of discovery and creativity. It challenges us to find the extraordinary in the seemingly ordinary, reminding us that beauty exists in the smallest of details. Through this unique fusion of art and science, we can share the marvels of the microscopic world in ways that inspire, educate, and illuminate.

Camera Setups for Microscopic Imaging

Navigating the realm of microscopic imaging requires not just a keen eye for detail but also the right tools for the job. Choosing the optimal camera setup for your microscope is akin to selecting the perfect lens through which to view and capture the wonders of the microscopic world. With an array of options available, from sophisticated digital microscope cameras to versatile DSLRs and even adaptable smartphone adapters, each choice brings its unique benefits to the table, tailored to meet various imaging requirements and aspirations.

Embarking on this selection journey, it's crucial to first acquaint yourself with the characteristics and capabilities of these different camera types. Digital microscope cameras, for example, are specifically designed for compatibility with microscopes, offering features such as live viewing on a computer screen, high-resolution images, and direct connectivity that can greatly enhance your microscopy experience. These cameras are often the go-to choice for those seeking precision and ease of use in scientific research or educational settings.

For enthusiasts looking to combine their love for photography with microscopy, DSLR cameras present a compelling option. With their large sensors and high megapixel counts, DSLRs can capture stunningly detailed images through the microscope. Though integrating a DSLR with a microscope might require additional adapters and a bit of setup, the results can be well worth the effort, producing images that not only serve scientific purposes but are also visually captivating.

In today's digital age, smartphone adapters have emerged as a surprisingly effective tool for microscopic imaging. Offering convenience and portability, these adapters allow you to mount your smartphone onto the microscope's eyepiece, enabling you to capture and share the beauty of the microscopic world instantly. While they may not match the image quality of dedicated cameras or DSLRs, smartphone adapters are an excellent way for beginners to dip their toes into the world of photomicrography without significant investment.

Selecting the right camera setup for microscopic imaging is about matching your specific needs, budget, and skill level with the right equipment. By understanding the strengths and limitations of each type of camera, you can make an informed choice that enhances your microscopy endeavors, allowing you to capture the unseen beauty of our world with clarity and precision. Remember, the best camera setup is the one that aligns with your goals and fuels your passion for discovery.

Utilizing Software for Microscopic Imaging

In the realm of microscopic imaging, capturing the perfect shot is only half the battle. The other half is where the magic of software comes into play, transforming raw images into captivating visuals and insightful data. Navigating through the wide array of software tools available can feel like unlocking a treasure chest filled with endless possibilities for enhancing and analyzing your microscopic adventures.

Embarking on this digital journey, you'll discover tools designed to stitch together multiple images into a single, high-resolution panorama, enabling you to view your specimen in its entirety with breathtaking detail. This technique is particularly useful for large samples that cannot be captured in one frame, ensuring you don't miss out on the bigger picture.

Color correction and enhancement tools open the doors to a world where the true vibrancy of your subjects comes to life. With a few clicks, you can adjust brightness, contrast, and saturation to reveal hidden details previously masked by uneven lighting or faint staining. These adjustments not only improve the aesthetic appeal of your images but also make it easier to identify specific features or patterns.

But the true power of software in microscopy lies in its ability to quantify what we see. Through image analysis functions, you can measure distances, areas, and volumes within your images, providing quantitative data to support your qualitative observations. This level of analysis can unveil relationships and patterns that are crucial for scientific research, adding a layer of objectivity to your microscopic explorations.

Advanced software tools also offer capabilities like 3D reconstruction, allowing you to build three-dimensional models from a series of 2D images. This immersive view can be pivotal in understanding the structure and spatial relationships within complex specimens, offering perspectives that flat images simply cannot provide.

Whether you're enhancing the visual quality of your images for presentation or diving deep into data analysis for research, software tools are your allies in unlocking the full potential of your microscopic investigations. By mastering these digital enhancements, you bring to light the unseen beauty and intricacies of the microscopic world, making every discovery a journey worth sharing.

Basic Principles of Image Analysis

Diving into the world of image analysis in microscopy is akin to unlocking a treasure trove of insights hidden within your images. It's here that the true potential of your microscopic endeavors comes to life, turning visuals into valuable data. As we embark on this exploration, let's demystify the foundational principles that pave the way for meaningful analysis.

At the heart of image analysis lies the concept of thresholding. Imagine it as setting a virtual gate that distinguishes the areas of interest in your image from the background. This selective focus allows you to isolate specific elements, making it easier to study their characteristics without the distraction of irrelevant details. It's a bit like choosing to listen to only one instrument in an orchestra to appreciate its unique contribution to the symphony.

Next up is segmentation, a step that further refines what was initiated by thresholding. If we continue with our musical analogy, segmentation is like isolating not just the sound of the instrument but also understanding the notes it plays. In microscopic terms, this means dividing the image into segments to analyze different structures or regions independently. Whether you're tracking the growth of cells or measuring the distribution of particles, segmentation is your guide to detailed analysis.

Feature extraction, another cornerstone principle, takes you deeper into the analytical journey. It's about identifying and quantifying the specific attributes of the segments you've isolated —think of it as recognizing the rhythm, pitch, and volume of our metaphorical instrument's notes. In microscopy, this could involve measuring the area, perimeter, or intensity of the structures you're studying, providing a quantitative basis for your observations.

Together, these principles form the backbone of image analysis in microscopy. By understanding and applying thresholding, segmentation, and feature extraction, you transform what might initially appear as merely beautiful images into rich sources of data. This process is not just about seeing but understanding, not just observing but discovering. As you master these principles, you'll find yourself not just looking at the microscopic world but truly comprehending its wonders.

Software Tools for Measurement and Enhancement

The journey through the microscopic world doesn't end at the click of the camera. With the right software tools, the realm of possibilities expands, allowing for both the enhancement of image quality and the precise measurement of intriguing details captured in your microscopic endeavors. Venturing into this digital domain arms you with the capability to not only observe but also to quantify and beautify the hidden marvels of the microcosm.

Embarking on this adventure, you'll find allies in software such as ImageJ and FIJI—powerhouses in the world of scientific image processing. These tools stand ready to assist in the meticulous measurement of objects within your images, from calculating areas and perimeters to assessing densities and intensities. This level of detail transforms your images from mere snapshots into valuable data sources, enabling you to conduct robust analyses and draw meaningful conclusions from your microscopic investigations.

But measurement is only one side of the coin. The art of microscopy is equally important, where enhancement software like Photoshop enters the picture. Here, the focus shifts to refining the visual aspects of your images, ensuring that every detail shines. Adjustments in brightness, contrast, and color saturation not only make your images more aesthetically pleasing but also can help to clarify structures and features, making them easier for both you and others to interpret.

Moreover, these tools open the door to creative storytelling through your images. By stitching multiple images into panoramas or employing techniques to increase depth of field, you can craft visuals that capture the imagination, inviting viewers on a journey through the microscopic landscape you've explored. This fusion of scientific precision and artistic expression elevates your work, allowing you to communicate the beauty and complexity of the microscopic world in ways that are engaging and accessible.

In harnessing the power of measurement and enhancement software, you become a master of your microscopic domain. Each image processed and every detail quantified brings you closer to unraveling the mysteries of the minuscule, bridging the gap between seeing and understanding, between observing and discovering. This digital toolkit is your companion on the path to sharing the wonders of the microscopic world, empowering you to illuminate the unseen in remarkable detail.

Sharing Your Microscopic Discoveries

Embarking on the final leg of our microscopy adventure leads us to one of the most exhilarating milestones: sharing the fruits of your labor with the world. Whether you've spent countless hours behind the lens capturing the unseen beauty of microscopic entities or meticulously analyzing each detail to unveil new insights, the journey doesn't truly end until your discoveries are brought into the light for others to see and learn from.

The act of sharing your microscopic findings is more than just a formality; it's a celebration of curiosity, dedication, and the relentless pursuit of knowledge. It bridges the gap between individual exploration and collective enlightenment, enabling your work to spark curiosity, fuel further research, and perhaps even revolutionize our understanding of the microscopic world.

Navigating the avenues for sharing can be as simple or as expansive as you choose. From presenting at academic and industry conferences, where the exchange of ideas fosters growth and collaboration, to publishing in esteemed scientific journals, your research has the potential to reach corners of the globe you may never visit. But let's not overlook the power of digital platforms and social media, where your images and insights can captivate an audience far beyond the scientific community, igniting a sense of wonder and appreciation for the microcosm among the general public.

Consider also the impact of crafting engaging educational content, whether it's through blog posts, video tutorials, or interactive webinars. By demystifying the complex techniques and sharing the stories behind your images, you make the microscopic world accessible and inspiring to budding scientists, curious learners, and the public at large.

Remember, each image shared is a beacon of knowledge, each finding a thread in the larger tapestry of scientific discovery. Your contributions, no matter how small they may seem, are vital pieces of a puzzle that helps us all to see the world in new and profound ways. So, take pride in your work, share it with enthusiasm, and watch as your microscopic discoveries ripple out, leaving a lasting impact on the world of science and beyond.

CHAPTER 9: APPLICATIONS OF MICROSCOPY

Exploring the Basics of Microscopy

Diving into the realm of the very small, microscopy offers us a fascinating window into a world unseen by the naked eye. But what exactly is this tool that acts as a bridge between us and the microscopic universe? At its heart, microscopy is the technical field that employs microscopes to view samples and objects that cannot be seen with the unaided eye. But it's not just about making things larger; it's about unveiling a whole new dimension of detail and complexity.

There are a variety of microscopes that cater to different purposes and offer different levels of magnification and resolution. Light microscopes, for instance, use visible light to illuminate and magnify objects, making them suitable for a broad range of applications from studying cell structures to observing the fine details of minerals. Electron microscopes, on the other hand, use a beam of electrons instead of light, offering much higher resolutions and allowing scientists to explore structures at the nano level, such as viruses or the detailed architecture of cell membranes.

Scanning probe microscopes break the mold entirely, not by using light or electrons to view samples, but by physically scanning a probe over a specimen to map out its surface structure. This method can reveal surface details at the atomic level, opening up exciting possibilities in materials science and nanotechnology.

Each type of microscope has opened new avenues for exploration and understanding in various fields. By enabling us to study the intricate details of cells, tissues, microorganisms, and materials, microscopy has not just expanded our view of the world—it has fundamentally transformed it. Through the lens of a microscope, we can begin to appreciate the complexity and beauty of the tiny engines that drive life and matter.

Biology and Medicine - Unveiling the Microscopic World

In the quest to understand the vast complexities of life and health, microscopy stands as an indispensable ally in biology and medicine. This fascinating tool brings into focus the unseen intricacies of the human body, from the dynamic dance of cells to the silent spread of diseases. Through the lens of a microscope, medical professionals and researchers gain critical insights that lead to groundbreaking advancements in treatment and diagnosis.

Delving into the cellular level, microscopy enables us to chart the map of life itself. Observing cells in their natural or altered states reveals the fundamental processes that sustain health or signal disease. Such detailed views are not just enlightening; they are transformative, offering clues to unraveling the mysteries of genetic disorders, cancer, and infectious diseases. The ability to pinpoint the characteristics of abnormal cells, for instance, guides pathologists in diagnosing conditions with greater accuracy and speed.

Beyond individual cells, microscopy plays a pivotal role in tissue analysis. This application is particularly crucial in understanding the architecture of tissues and organs, how they interact, and what happens when disease disrupts their harmony. It's a window into the effects of diseases at a macro level, providing a comprehensive view that enriches our understanding of human anatomy and physiology.

Moreover, the power of microscopy in medicine extends to the front lines of fighting infectious diseases. It helps in identifying the culprits behind infections—be it bacteria, viruses, or fungi—allowing for targeted treatments and the development of vaccines. This aspect of microscopy not only aids in current medical practice but also in epidemiological surveillance and the prevention of future outbreaks.

In essence, the application of microscopy in biology and medicine is a journey of discovery, one that continuously uncovers the wonders and woes of the microscopic world. It is a testament to the relentless human pursuit of knowledge, a pursuit that has led to life-saving medical breakthroughs and a deeper appreciation of the biological tapestry that weaves the story of life.

Cell and Tissue Analysis - Understanding the Building Blocks

Cell and tissue analysis stands as a cornerstone of microscopy in both biology and medicine, inviting us to delve deeper into the mysteries of life at the microscopic level. Through the lens of a microscope, scientists and researchers embark on an insightful journey into the very fabric of our existence—cells and tissues. These tiny structures, invisible to the naked eye, are the fundamental components of all living organisms, and understanding them is crucial for unlocking the secrets of how life operates.

Microscopy, in its various forms, equips us with the ability to not only see but also analyze the intricate details of cells and tissues. This analysis sheds light on how cells function, interact, and organize themselves into the complex tissues that form our organs and systems. By observing these building blocks, scientists can identify the hallmarks of health and disease, leading to better diagnostic techniques and treatments. For instance, in cancer research, microscopy allows for the examination of tumor samples at a cellular level, helping to identify cancerous cells by their appearance and behavior. This is pivotal in determining the aggressiveness of a cancer and tailoring treatment plans to combat it effectively.

Furthermore, cell and tissue analysis is instrumental in the field of regenerative medicine, where understanding cell dynamics and tissue architecture paves the way for innovations like organ regeneration and tissue engineering. By scrutinizing cells and tissues through microscopes, researchers can observe the processes of cell growth, division, and differentiation, uncovering clues that could lead to breakthroughs in healing and regenerative therapies.

In essence, the analysis of cells and tissues through microscopy offers a window into the biological intricacies that sustain life. It enables scientists to explore the unseen, providing insights that have profound implications for medicine, genetics, and beyond. As we continue to harness the power of microscopy, our journey into the microscopic world promises to reveal even more about the complex interplay of cells and tissues, furthering our understanding of the building blocks of life.

Microorganism Identification - Exploring the Invisible

Diving into the world of tiny, yet incredibly influential beings, microorganism identification through microscopy is nothing short of a microscopic adventure. This fascinating process opens up the unseen world of bacteria, viruses, fungi, and other microorganisms, laying bare their secrets and showing us how they shape our world, influence our health, and interact with the environment. With the aid of microscopes, researchers and scientists embark on a quest to categorize and understand these micro inhabitants, each discovery contributing to a vast library of knowledge that spans several scientific disciplines.

Microscopy, in this context, is the detective's tool of choice, allowing us to observe the unique shapes, structures, and movements that distinguish one microorganism from another. The level of detail revealed can be astonishing—from the complex cell walls of bacteria to the intricate capsids of viruses. These observations are critical not only for academic curiosity but also for practical applications such as diagnosing diseases, improving public health strategies, and developing new treatments and vaccines.

One of the most thrilling aspects of microorganism identification is the detective work involved in linking specific strains to their roles in health and disease. For example, by identifying the specific bacteria responsible for an infection, clinicians can prescribe the most effective antibiotics, significantly improving patient outcomes. Likewise, understanding the behavior of viruses can lead to more effective vaccines and antiviral therapies, showcasing the direct impact of microscopy on our well-being.

Furthermore, the study of microorganisms extends beyond the realm of human health, touching on environmental issues, food safety, and even the search for new forms of energy. Through the lens of a microscope, we gain insights into microbial ecosystems, discover new species, and unravel the complex interactions that sustain life on Earth.

As we continue to explore this invisible world, the contributions of microscopy to microorganism identification promise to keep pushing the boundaries of what we know, enabling us to face future challenges with better-informed strategies and solutions.

Materials Science - A Closer Look at What Things Are Made Of

Dive into the microscopic universe of materials science, where the very fabric of our everyday objects is examined and reimagined. This fascinating realm utilizes microscopy to peel back the layers and unveil the secrets held within materials, from the metals that fortify our bridges to the polymers that encase our electronics. Imagine being able to zoom in, way in, to see the atoms and molecules that constitute these materials. That's exactly what scientists do with the help of cutting-edge microscopes, such as scanning electron microscopes (SEM) and atomic force microscopes (AFM).

In materials science, the journey through a microscope lens is akin to an explorer charting unknown territories. Each material—be it a slice of silicon or a piece of polymer—tells a story through its microscopic structure. By analyzing these structures, scientists can understand why certain materials behave the way they do under different conditions. For instance, why does one type of steel withstand corrosion better than another? How can we create a polymer that's both lightweight and incredibly strong? Microscopy lays the groundwork for answering these pivotal questions.

But the applications don't stop at mere observation. Armed with insights gleaned from microscopic analysis, researchers are pushing the boundaries of innovation. They're engineering materials with tailor-made properties—imagine metals that can heal their own cracks or glasses that are nearly impossible to shatter. The possibilities are as vast as the microscopic world itself.

This journey into the heart of materials doesn't just enhance our understanding; it directly impacts the development of new technologies and improvements in industries ranging from aerospace to electronics. With each microscopic examination, the field of materials science is crafting the building blocks of tomorrow's world, piece by tiny piece.

Studying Metals and Crystals - Revealing Their Inner Secrets

When we venture into the microscopic examination of metals and crystals, we're embarking on a thrilling expedition to uncover their hidden marvels. This journey enables scientists to peel back the layers and dive deep into the atomic and molecular makeup of these materials, shedding light on their unique properties and behaviors. With the aid of sophisticated techniques like electron microscopy and X-ray diffraction, the very blueprint of metals and crystals is revealed.

This intricate exploration is not just for the sake of curiosity. Understanding the atomic structure of these materials is pivotal for numerous applications, from crafting stronger alloys to designing more efficient semiconductors. Imagine a world where metals can resist extreme temperatures or where crystals are engineered to have precise electronic properties. The potential is boundless, and it all starts with the microscopic study of their inner workings.

Microscopy's role in this field is akin to giving scientists superpowers. With these tools, they can see the arrangement of atoms and observe the defects and irregularities that might affect a material's strength or electrical properties. This microscopic insight is crucial for innovation. For instance, by understanding how atoms are arranged in a metal, researchers can develop new materials that are lighter, stronger, or more flexible than anything currently available.

Moreover, the study of metals and crystals at the microscopic level also plays a critical role in addressing real-world challenges. From creating more durable building materials to advancing the technology in our smartphones, the secrets unlocked by microscopy fuel the advancements that shape our modern life. It's a fascinating journey into the small that makes a big impact on our understanding and manipulation of the materials that surround us.

Nanotechnology - The Frontier of the Tiny

Welcome to the riveting frontier of nanotechnology, where the marvels of the microscopic world unfold in extraordinary detail through the lens of advanced microscopy. Here, scientists don the hats of explorers, delving into the realm of the unimaginably small—the nanoscale. It's a place where materials behave in ways that defy the physics of the bulk world, opening up a treasure trove of possibilities.

At the core of this exploration is microscopy, a key that unlocks the door to viewing and manipulating matter at the atomic and molecular levels. Techniques such as transmission electron microscopy (TEM) and scanning probe microscopy (SPM) serve as the microscopes' superpowers, allowing us to not only see but also interact with this nano world. These powerful tools enable scientists to observe the structure, properties, and behavior of nanoparticles and nanostructures with unprecedented clarity.

Imagine being able to tailor materials with precision, crafting nanoparticles that can target and destroy cancer cells, or designing nanostructures that clean pollutants from water with efficiency we've never seen before. This is the potential that nanotechnology holds, and it's microscopy that's lighting the way. Through meticulous observation and manipulation of nanomaterials, researchers are developing innovative solutions to some of the world's most pressing challenges, from healthcare to environmental protection.

The journey through nanotechnology is filled with discovery and innovation, driven by the insights gained from microscopy. As we continue to push the boundaries of what's possible at the nanoscale, the impact of these tiny wonders on our lives promises to be as vast as the universe itself, and it's an adventure that's just getting started.

Environmental Science - Investigating Our Natural World

Venturing into the microscopic realms of our planet, environmental science harnesses the power of microscopy to unveil the intricate details of Earth's natural systems. This vital tool equips scientists with the ability to closely examine the smallest constituents of our environment, offering a magnified glimpse into the workings of ecosystems that are often invisible to the eye. From the bustling life within a single drop of pond water to the complex interplay of particles in a handful of soil, microscopy reveals the unseen forces that sustain and affect our natural world.

In this enlightening journey, environmental scientists use microscopes to scrutinize various samples, such as soil and water, shedding light on the microorganisms, pollutants, and minerals contained within. This microscopic exploration is not just about observation; it's a means to understand the profound impact of human activities on the environment and to forge paths toward sustainable living and conservation efforts. By identifying the presence of specific bacteria in water or detecting pollutants in soil, microscopy aids in assessing the health of ecosystems and pinpointing sources of contamination.

Moreover, the study of microscopic life through environmental microscopy is pivotal in comprehending biological processes, such as decomposition and nutrient cycling, that are critical to ecosystem function. It helps in mapping out the biodiversity of microscopic life, contributing valuable insights into how these tiny entities interact within ecosystems and respond to environmental changes.

Through the lens of a microscope, environmental science becomes a dynamic field of discovery and insight, where every grain of soil and drop of water tells a story of balance, disruption, and coexistence. It's a microscopic expedition that illuminates the complexities of our natural world, emphasizing how even the smallest creatures and particles play significant roles in the tapestry of life on Earth.

Soil and Water Analysis - Microscopy at the Earth's Surface

Embarking on a microscopic exploration of soil and water may sound like a dive into the depths of our planet, but it's actually a journey that unfolds right at the Earth's surface. Through the eye of a microscope, the intricate world of soil particles, minerals, and the teeming life within water bodies is brought into stunning detail. This analysis isn't just for the curious-minded; it plays a pivotal role in understanding the health and vibrancy of our environment.

In the study of soil, microscopy unveils a labyrinth of particles and microorganisms, each contributing to the soil's unique structure and fertility. Scientists use microscopes to delve into the composition of soil, revealing how different types of soil support various plant life and ecosystems. Similarly, water analysis through microscopy introduces us to a bustling microworld. From algae that oxygenate our waters to bacteria that recycle nutrients, the microscopic inhabitants of water bodies play crucial roles in maintaining ecological balance.

But there's more to soil and water analysis than meets the eye. By scrutinizing these samples under a microscope, environmental scientists can detect signs of pollution or contamination long before they become visible or cause widespread harm. This allows for timely interventions to protect our water sources and ensure the soil remains fertile and healthy.

Microscopy, in the context of soil and water analysis, is a gateway to uncovering the unseen forces that shape our natural world. It's a tool that not only magnifies our view but also enhances our understanding, helping us make informed decisions to safeguard our planet for future generations.

Microbial Ecology - Uncovering Microscopic Ecosystems

Dive into the hidden realm of microbial ecology with us, where every droplet of water, handful of soil, or breath of air teems with an astonishing array of microorganisms. Through the lens of a microscope, we embark on an incredible journey to discover these tiny but mighty inhabitants that play essential roles in Earth's ecosystems. Microscopy serves as our trusty guide, revealing the diversity and behaviors of bacteria, fungi, and other microorganisms in vivid detail.This exploration into microscopic ecosystems isn't just about marveling at the tiny life forms. It's about understanding the critical functions these microorganisms perform - from breaking down organic matter and recycling nutrients, to influencing the climate and contributing to plant health. Each microbe has a story, a purpose in the tapestry of life that we're only beginning to understand.By examining microbial communities across different environments, scientists uncover how these organisms interact with each other and with their surroundings. It's a dynamic world down there, with microbes competing, collaborating, and contributing to the balance of natural systems. Microscopy in microbial ecology also shines a light on the unseen influences of human activity on microbial communities. As we explore this microscopic world, we gain insights into how to protect and sustain the delicate ecosystems that these microorganisms support. It's a journey of discovery, filled with surprises at every turn, and it underscores the importance of even the smallest creatures in maintaining the health of our planet.Join us as we continue to uncover the secrets of microbial ecology, one microscopic ecosystem at a time, and celebrate the unseen heroes of Earth's natural systems.

CHAPTER 10: MICROSCOPY UNVEILED, EXPLORING ADVANCED TECHNIQUES

The Illuminating World of Fluorescence Microscopy

Dive into the vibrant realm of Fluorescence Microscopy, where the unseen becomes visible in a spectacular display of colors and lights. This advanced microscopy technique stands out as it employs fluorescent dyes and markers, making it possible to illuminate specific components within a specimen with extraordinary clarity and resolution. Imagine being able to light up just the parts of a cell you are interested in, while the rest fades into the background, allowing for precise observation and study. This is the magic of fluorescence microscopy.

At its core, fluorescence microscopy is not just about creating visually stunning images; it's a doorway to understanding the complexities of cellular structures, interactions, and functions. By tagging molecules of interest with fluorescent dyes that glow under specific wavelengths of light, scientists can trace the pathways of these molecules, observe how cells divide or interact, and detect the presence of particular proteins. This capability has invaluable applications across a broad range of scientific fields, including biology, where it's essential for understanding diseases at a cellular level, and in materials science, for analyzing the properties of synthetic and natural materials.

The process begins with the careful selection of fluorescent dyes and markers, which are chosen based on their ability to bind to specific targets within the specimen. Whether it's DNA, various proteins, or cell organelles, these markers light up when exposed to light at certain wavelengths, emitting fluorescence. This emitted light is then captured to produce clear, detailed images of the sample, revealing insights that would be impossible to gain with traditional light microscopy.

One of the most captivating aspects of fluorescence microscopy is its versatility. It can be combined with other microscopy techniques, such as live-cell imaging, to watch cells in action in real time, providing a dynamic view of biological processes as they happen. Moreover, the technique has been pivotal in the advancement of medical research, aiding in the diagnosis and understanding of diseases by allowing scientists to pinpoint and visualize the distribution of specific molecules within cells and tissues.

Embarking on a journey with fluorescence microscopy opens a window to a world filled with wonder, where the hidden beauty and complexity of the microscopic world come to light. As we continue to explore and push the boundaries of what can be seen and understood, fluorescence, with its colorful glow, remains an indispensable ally in the quest for knowledge.

Fluorescent Dyes and Markers: Lighting the Way

At the heart of the spectacular visuals produced by fluorescence microscopy are the fluorescent dyes and markers, the unsung heroes that illuminate the hidden intricacies of cells and molecules. These specially engineered molecules are not just any ordinary dyes; they are carefully selected based on their ability to attach themselves to specific targets within a sample, such as proteins, DNA, or cell membranes. This specificity allows scientists and researchers to spotlight particular structures or functions within a cell, making it possible to observe and analyze them in great detail.

Choosing the right fluorescent dye or marker is akin to selecting the correct key for a lock. Each dye is designed to emit light at a specific wavelength when excited by light of a certain wavelength, which means that by using a combination of different dyes, researchers can simultaneously highlight multiple targets in a multitude of colors. This ability to label different structures or molecules with different colors is incredibly powerful, offering a multi-faceted view of the sample's biological processes.

Moreover, the development of advanced fluorescent dyes and markers continues to expand the capabilities of fluorescence microscopy. Innovations in dye chemistry have led to the creation of more stable, brighter, and more photostable dyes. These improvements enhance the quality and duration of imaging, allowing for longer observations of dynamic processes within living cells without significant photobleaching or damage to the samples.

In practice, applying these dyes and markers requires meticulous planning and precision. The process often begins with the fixation and permeabilization of the sample, which prepares the cells or tissues to accept the fluorescent markers. Following this, the dyes are applied to the sample, where they bind to their target structures. After washing away any unbound dye, the sample is ready for imaging. The result is a brilliantly colored window into the microscopic world, offering insights that are not just visually stunning but scientifically valuable.

As we journey further into the exploration of cellular and

molecular biology, fluorescent dyes and markers continue to play a pivotal role. They not only allow us to see the unseen but also to understand the complex dance of life at a microscopic level. Their ongoing development and refinement promise even greater discoveries and insights, lighting the way for future research and innovation in microscopy.

Unraveling the Layers with Confocal Microscopy

Unraveling the mysteries of the microscopic world just got a whole lot easier with confocal microscopy. This ingenious technique enhances the clarity and resolution of fluorescence images in a way that feels almost like magic. Imagine peeling away the layers of an onion, focusing intently on one slice at a time without losing sight of the whole picture. That's the essence of confocal microscopy – it uses a clever pinhole aperture to shine a light on one specific plane of your sample at a time, cutting through the haze of out-of-focus light. This selective illumination brings forth sharp, clear images that are rich in detail, making it a go-to method for examining complex, thick samples or three-dimensional structures with unparalleled precision.

What sets confocal microscopy apart is its ability to provide optical sectioning capabilities, a feature that traditional fluorescence microscopy can't offer. Think of it as having the superpower to focus on each layer of your sample, one at a time, without physically cutting it. This is particularly useful when you're delving into the intricate details of cells and tissues or exploring the spatial arrangements within materials. The ability to adjust the focus to different depths within the sample allows for the construction of three-dimensional images, layer by layer, providing insights into the architecture and organization of microscopic structures.

The real magic happens in the dance between the fluorescent dyes or markers, introduced earlier, and the confocal microscopy technique. When these fluorescently tagged samples are illuminated at precise wavelengths, only the fluorophores in the focal plane emit light that passes through the pinhole. This selective detection ensures that only the light from the plane of interest is captured, dramatically enhancing image resolution and contrast. The result? Vibrant, detailed images that bring the unseen world into vivid focus, offering a window into the microscopic realm that's both mesmerizing and informative.

Embarking on an exploration with confocal microscopy not only enriches our understanding of the microscopic world but also opens up new avenues for scientific discovery and innovation. By enabling detailed studies of cellular structures, tissue sections, and even non-biological specimens, confocal microscopy serves as a powerful tool in the arsenal of modern science, paving the way for advancements across diverse fields. Whether it's unraveling the complexities of biological systems or unveiling the secrets of materials at the microscale, confocal microscopy continues to illuminate our journey into the microscopic universe.

The Art and Science of Sample Preparation in Confocal Imaging

Diving into the fascinating world of confocal imaging requires not just a keen eye for detail but also a meticulous approach to sample preparation. It's an art and science that, when mastered, unlocks the full potential of confocal microscopy, allowing us to capture stunning images that shed light on the mysteries of the microscopic realm.

Sample preparation for confocal microscopy is like setting the stage for a grand performance. Each step, from fixing and staining to mounting on a slide, is performed with precision and care to ensure that the star performers, our samples, can shine under the spotlight. It begins with the fixation of the sample, a critical step that preserves the natural structure and composition of the specimen. This is usually followed by staining with fluorescent dyes or markers, a process that highlights the specific features or components we are interested in observing.

The choice of fluorescent dyes and markers is crucial. As we learned earlier, these are not just ordinary dyes; they are the keys to revealing the hidden beauty of our samples. They bind to specific targets, illuminating them against the backdrop, and allowing us to peer into the biological processes at play with incredible clarity.

But the artistry doesn't end with staining. Mounting the sample on a slide is equally important. This step involves carefully placing the specimen in a medium that supports and preserves its structure during imaging. The thickness and orientation of the sample are also vital considerations, as they can greatly influence the quality of the resulting images. Too thick, and the sample may not be fully penetrated by light; too thin, and vital details might be lost.

Adjusting the imaging parameters is the final touch that brings the masterpiece to life. By fine-tuning the settings on the confocal microscope, we can capture images that are not only visually striking but also rich in scientific value. This includes adjusting the pinhole size, laser intensity, and filter settings to optimize the fluorescence signal and achieve high-resolution images.

In the world of confocal imaging, every detail matters. From the initial fixation to the final adjustments on the microscope, each step is a testament to the careful planning and precision that goes into preparing samples. It's a delicate balance between art and science, where the goal is not just to observe but to reveal the wonders of the microscopic world in all their glory.

Electron Microscopy: Revealing the Ultra-Small

Step into the world of Electron Microscopy, where the ultra-small is unveiled in unprecedented detail. This powerful imaging technique propels us into the nanometer-scale universe, revealing the intricate textures and structures invisible to the naked eye and other forms of microscopy. Electron microscopy stands as a titan among tools for scientists eager to explore the minutiae of materials science, biology, and beyond.

Electron microscopy isn't a one-size-fits-all approach; it encompasses two primary types that cater to different scientific cravings. Transmission Electron Microscopy (TEM) and Scanning Electron Microscopy (SEM) serve as the gateways to understanding the inner and outer worlds of our samples. TEM takes you on a journey through the inner landscapes of specimens, offering a glimpse into the very essence of cells and nanomaterials with astonishing resolution. Imagine being able to see the detailed architecture of a virus or the precise arrangement of atoms within a material. That's the prowess of TEM.

On the flip side, SEM lets us roam the surface of samples, capturing their topography and texture with breathtaking depth. The beauty of SEM lies in its ability to provide a three-dimensional perspective, bringing to life the surface characteristics of everything from pollen grains to metal alloys. It's like having the power to explore alien landscapes, except these worlds exist on the tips of our fingers and in the air we breathe.

Preparing samples for electron microscopy is an art form, requiring a blend of precision and understanding. The samples must undergo a meticulous preparation process that involves fixing, dehydrating, and embedding, ensuring they are primed for the spotlight under the electron beam. Each step is critical in preserving the natural state and avoiding artifacts that might distort our view into this tiny universe.

As we embrace electron microscopy, we're not just observing the unseen—we're unlocking the mysteries that lie within the very fabric of the material and biological worlds. This journey into the ultra-small is not merely about seeing but understanding the building blocks of nature at the most fundamental level. With electron microscopy, we're equipped to venture deeper into the microscopic frontier, expanding our knowledge and feeding our curiosity about the world around us.

TEM vs. SEM: A Comparative Look

Embarking on a journey into the microscopic realm requires a guide, and in the vast universe of electron microscopy, two heroes emerge to lead the way: Transmission Electron Microscopy (TEM) and Scanning Electron Microscopy (SEM). Each of these advanced techniques offers a unique perspective on the world at the nanoscale, but how do you know which path to take? Let's dive into a friendly and informative comparison to help navigate these waters.

TEM, the seasoned voyager of the microscopic landscape, provides an unparalleled view into the internal composition of specimens. It's like being shrunk down and walking through the inner corridors of cells and materials, witnessing the atomic structure with breathtaking detail. This technique is perfect for when your quest demands a deep dive into the very essence of a sample, to uncover secrets hidden beyond the reach of visible light. However, preparing for this journey requires careful planning. Samples must be sliced incredibly thin, often to the point of being transparent to electrons, a process that requires skill and patience.

On the other side, SEM acts as a rugged explorer, mapping the surface of samples with astonishing clarity and depth. Think of it as hiking over the rugged terrain of samples, where every ridge, valley, and texture is brought to life in exquisite detail. SEM's strength lies in its ability to produce three-dimensional images that give a lifelike representation of the sample's exterior. This technique is less demanding in terms of sample preparation compared to TEM, making it a more accessible route for those looking to capture the surface stories of their specimens.

Choosing between TEM and SEM boils down to the nature of the discovery you seek. Are you aiming to unravel the internal mysteries, diving deep into the cellular or material structure? TEM is your go-to. Or are you more interested in exploring the surface, seeking a comprehensive view of the topography and texture? SEM will serve you well.

Remember, each technique shines its own light on the microscopic world, offering insights and wonders unique to its path. The choice between TEM and SEM is not just a technical decision; it's a strategic step in your journey of exploration and discovery in the vast and intricate universe of microscopy.

Preparing Samples for the Electron Microscope

Embarking on the adventure of electron microscopy, whether you're leaning towards the insightful depths of TEM or the captivating surfaces revealed by SEM, requires a bit of groundwork to get your specimens ready for their close-up. Think of it as prepping a star before they step out on stage under the bright lights; every detail must be meticulously taken care of to ensure a stellar performance.

The journey of sample preparation is a delicate dance that begins with fixing the specimen. This initial step is all about preserving the sample's natural state, almost like capturing a moment in time. By using chemicals or freezing methods, we essentially halt the biological processes, preventing any further changes or degradation. It's a crucial step because it maintains the integrity and fine structures of the specimen, ensuring that what we see under the microscope is as true to life as possible.

Next up, we dive into the dehydration process. Here, water is carefully removed from the sample because, in the vacuum environment of an electron microscope, any remaining moisture could vaporize and damage the delicate structures we're so eager to observe. This step often involves a series of alcohol or acetone washes, each one increasing in concentration to gently coax the water out.

After our specimen is dry, we move on to embedding. This step is akin to setting our star in the perfect backdrop. The sample is placed in a medium that will harden, supporting and preserving its structure. For TEM, this usually means embedding it in resin that will be sectioned into ultra-thin slices. SEM samples might also be coated with a conductive material to help them deal with the electron beam they're about to face.

The final act in our preparation saga is the sectioning for TEM, where samples are cut into incredibly thin slices, so thin that electrons can pass through them. This requires precision and a steady hand, as the slices need to be just the right thickness to reveal those inner secrets without obscuring the view.

Each of these steps, from fixing to sectioning, is a testament to the care and precision that goes into electron microscopy. It's a process that marries the art of meticulous preparation with the science of imaging, setting the stage for discoveries that can change the way we understand the world on an almost infinitesimal scale.

Innovations in Microscopy: What's Next?

The journey through the microscopic universe is on the brink of an exciting era, fueled by remarkable technological breakthroughs. As we peer into the future, the landscape of microscopy is set to transform, unlocking dimensions and details previously beyond our grasp. Super-resolution microscopy is at the forefront of this revolution, shattering the limits imposed by the diffraction of light to capture images with stunning precision. This game-changer allows scientists to visualize structures at the nanoscale, bringing into focus the intricate details of molecules and cells in ways that were once thought impossible.

Equally groundbreaking, multi-photon microscopy is making waves by enabling deeper penetration into biological samples with minimal damage, offering a glimpse into the dynamic processes of living cells in their natural state. This technique relies on the simultaneous absorption of two or more photons, providing an exquisite level of detail and reducing phototoxicity, a boon for long-term live-cell observations.

Not to be outdone, correlative microscopy is merging the strengths of diverse imaging methods, such as fluorescence and electron microscopy, to provide a comprehensive view of samples. This synergy allows researchers to correlate functional insights with ultrastructural details, offering a holistic understanding of biological phenomena.

As these innovative techniques continue to evolve, they promise to open new vistas of discovery. The realm of the incredibly small is becoming increasingly accessible, heralding a new chapter in our quest to unravel the mysteries of the microscopic world. The future of microscopy is bright, and its potential is as vast as the universe it helps us explore.

CHAPTER 11: MASTERING MICROSCOPY: TROUBLESHOOTING COMMON ISSUES

Introduction to Microscopy Troubleshooting

Embarking on the journey of microscopy troubleshooting can seem daunting at first, but fear not! With a little patience and some basic know-how, you'll be navigating common issues with ease. Think of your microscope as a trusty companion in your scientific explorations—a companion that, just like any good friend, may sometimes need a bit of help to perform its best. Understanding the nuts and bolts of how your microscope operates is the first step in becoming a troubleshooting pro.

Microscopes are fascinating devices, employing a symphony of lenses, light, and mechanics to unveil the hidden details of the microscopic world. Each component, from the eyepiece to the objective lenses and the illumination system, plays a crucial role in capturing crisp, clear images. However, even a slight misstep in handling or setup can lead to common issues such as blurry images or inadequate lighting, which we'll tackle in more detail.

Troubleshooting is much like detective work. It involves observing the problem, understanding the potential causes, and methodically testing solutions until the issue is resolved. For instance, if you're faced with a blurry image, you'll want to check everything from the focus to the cleanliness of the lenses. Insufficient lighting, on the other hand, calls for an examination of your microscope's illumination settings and, perhaps, the exploration of additional light sources.

But where to start? First, familiarize yourself with the basic components of your microscope and their functions. This knowledge will empower you to identify which part may be causing the issue. Next, approach troubleshooting with a calm and systematic mindset. Remember, the goal is not just to fix the problem at hand, but to understand why it happened in the first place. This approach will not only resolve the current issue but also help you prevent similar problems in the future.

In this informative and friendly guide, we'll walk you through the steps to diagnose and solve the most common microscopy hitches. From adjusting your focus to perfecting your lighting, we're here to help you enhance your microscopy skills. So, let's dive into the fascinating world of microscopy troubleshooting together, with the confidence that you have all the tools and knowledge needed to tackle any challenge that comes your way.

Dealing with Blurry Images

Navigating through the hurdle of blurry images in microscopy can be quite the puzzle, but it's a challenge that's often easily solved with a few careful steps. Blurry images can dampen the excitement of discovery under the microscope, but don't worry, we're here to guide you through some straightforward solutions that can bring the microscopic world back into sharp focus.

First off, if you encounter a blurry image, the immediate action is to check your focus. Sometimes, the solution is as simple as fine-tuning the focus knob. This adjustment often requires a bit of patience and precision to find the sweet spot where your specimen snaps into clear view. Remember, the key here is gentle, incremental adjustments; rushing this process can overshoot the point of perfect focus.

If fiddling with the focus knob doesn't clear up the image, it's time to examine the cleanliness of your lenses. A common culprit behind blurry images is smudges, dust, or fingerprints on the microscope's lenses. Carefully cleaning the lens with a soft, lint-free cloth, preferably one that's designed for optical instruments, can make a world of difference. It's like cleaning your glasses; you'll be surprised at how much clearer your view is afterward.

Another factor to consider is the alignment of your microscope's components. Misalignment can lead to images that are persistently out of focus. Ensuring that the eyepiece, objective lenses, and other optical elements are correctly aligned is crucial for obtaining crisp images. This might sound daunting, but it's often a matter of consulting your microscope's manual for alignment instructions. Some microscopes have user-friendly alignment features, making this task more approachable than it might initially seem.

While addressing blurry images, it's also important to consider the condition of the specimen itself. A specimen that's too thick or not properly mounted can result in poor image quality. Make sure your specimen is prepared correctly for observation. This may involve adjusting its thickness or ensuring it's properly centered and secured on the slide.

In summary, tackling blurry images in microscopy involves a combination of checking focus, cleaning lenses, ensuring proper alignment, and verifying specimen preparation. With these steps, you're well-equipped to resolve one of the most common challenges in microscopy, paving the way for clearer, more detailed observations of the tiny wonders that await you.

Overcoming Insufficient Lighting

When tackling the challenge of insufficient lighting in microscopy, you're addressing one of the pivotal aspects that can make or break your microscopic examination. Achieving the right balance of illumination is not just about seeing your specimen; it's about revealing the vivid details that bring your observations to life. Let's walk through some illuminating tips to enhance your microscopy experience.

The first checkpoint in your quest for better lighting is the microscope's own light source. Many modern microscopes come equipped with adjustable light sources, allowing you to fine-tune the brightness to suit your needs. Experiment with different levels of illumination to find what best highlights the features of your specimen. Remember, more light isn't always the answer; too much brightness can wash out the details you're trying to observe.

If adjusting the built-in light source doesn't quite cut it, consider the role of external lighting solutions. The addition of an auxiliary light, such as an adjustable LED lamp, can provide the extra illumination needed for clearer, more detailed images. Positioning an external light source correctly can minimize shadows and evenly distribute light across your specimen, giving you a better viewing experience.

Another aspect often overlooked is the type of lighting. Microscopes generally use either transmitted light (light passing through the specimen) or reflected light (light reflecting off the specimen). Your choice between these two can dramatically affect the visibility of your specimen. Transmitted light is ideal for transparent or semi-transparent specimens, while reflected light works best for opaque specimens. Understanding the nature of your specimen and experimenting with these lighting techniques can unlock new levels of detail in your observations.

Don't forget about the role of condensers and diaphragms in managing light. These components help focus and control the amount of light that reaches your specimen. Adjusting the condenser and diaphragm can improve contrast and clarity, making it easier to discern the subtle features of your specimen.

In the journey to overcome insufficient lighting, patience and experimentation are your best tools. Each specimen is unique, and finding the perfect lighting setup requires a bit of trial and error. By taking the time to explore different lighting options and adjustments, you're not just solving a problem; you're enhancing your ability to observe the microscopic world in all its glory.

Essential Microscope Maintenance

Maintaining your microscope isn't just about keeping it clean; it's about ensuring that every exploration into the microscopic world is as clear and enlightening as the first. Think of your microscope as a high-performance vehicle. Just as you would regularly service your car to keep it running smoothly, your microscope requires regular maintenance to perform at its best.

The cornerstone of effective microscope maintenance is a routine cleaning regimen. Dust, oils, and other contaminants can accumulate on the lens surfaces and other components, significantly impacting the quality of your images. Gentle is the operative word here. Use a soft, lint-free cloth designed for optical equipment, and if necessary, a mild, alcohol-based cleaning solution specifically recommended for microscopes to carefully clean the exterior surfaces and lenses. This simple step can vastly improve image clarity and prevent the common pitfall of blurry, unclear observations.

But there's more to maintenance than just keeping things clean. Regular checks and adjustments are crucial to keep your microscope aligned and calibrated. Over time, the mechanical components of your microscope, such as the focus knobs and stage adjustments, can become misaligned due to regular wear and tear. Periodically, it's important to check that these components are moving smoothly and accurately. If something feels off, refer to your microscope's manual for guidance on how to make adjustments or when to seek professional servicing.

Another aspect of maintenance that's often overlooked is the proper storage of your microscope. When not in use, cover your microscope with a dust cover or store it in a cabinet to protect it from airborne particles. Additionally, ensure your microscope is kept in a dry, temperature-controlled environment to avoid damage from humidity or extreme temperatures.

Lastly, remember that maintenance is not just a solo task. Involve your peers or team members in the care regimen. Sharing the responsibility ensures everyone benefits from a well-maintained instrument and contributes to the longevity of the microscope. Encourage a culture of care and mindfulness when using shared equipment, reinforcing the idea that a well-maintained microscope is a shared asset that can elevate the quality of everyone's research.

By incorporating these maintenance practices into your routine, you're not just caring for a piece of equipment; you're safeguarding your window into the microscopic world. Regular maintenance keeps that window clear, allowing you to continue unveiling the tiny wonders that enhance our understanding of the world around us.

The Art of Cleaning and Care

Delving into the art of cleaning and care for your microscope, it's essential to approach this task with the finesse and attention it deserves. Think of your microscope not just as a tool, but as a trusted partner in your journey of discovery. Proper care ensures that this partnership remains productive and enduring. The process of cleaning your microscope goes beyond mere aesthetics; it directly impacts the clarity and quality of the images you obtain.

To begin, always ensure that your hands are clean before handling any part of the microscope. Natural oils from your skin can easily transfer to optical surfaces and degrade the quality of your observations. When it's time to clean the lenses, which are the heart of your microscope's optical system, a specific technique is recommended. Opt for a lens cleaning solution approved for optical surfaces, applying it sparingly to a soft, lint-free cloth or lens paper. Gently wipe the lens in a circular motion, starting from the center and moving outward. This method helps prevent streaks and ensures that any particulates are not dragged across the lens surface, which could cause scratches.

For the body of the microscope and non-optical surfaces, a slightly dampened cloth with water or a mild cleaning solution can be used. Avoid spraying any solution directly onto the microscope, as excess moisture can seep into internal components, leading to potential damage. After wiping, pass over the same areas with a dry cloth to remove any residual moisture.

Dust can be a persistent adversary in maintaining your microscope. Utilize a soft, fine-bristled brush or compressed air to gently remove dust from hard-to-reach places. However, be cautious with compressed air, as the force can potentially displace smaller components or drive dust deeper into the microscope.

When not in use, shielding your microscope with a dust cover or storing it in a cabinet is an effective way to minimize dust accumulation and exposure to environmental factors. Such proactive measures extend the intervals between deep cleanings and help maintain the instrument's performance.

Remember, the objective of cleaning and care is not just about preservation but also about enhancing your microscope's performance. By incorporating these practices into your routine, you ensure that your microscope remains a clear window to the microscopic world, ready to reveal its secrets whenever you are.

Calibration and Alignment for Optimal Performance

Diving into the world of microscopy, one quickly realizes that the sharpness of an image isn't just about having a keen eye; it's about the precision with which your microscope is calibrated and aligned. Think of calibration and alignment as the fine-tuning of a musical instrument. Just as a perfectly tuned guitar produces harmonious melodies, a well-calibrated and aligned microscope brings the microscopic world into clear, precise focus.

Embarking on this journey, let's first explore calibration. Calibration is the process of verifying and adjusting the accuracy of your microscope's settings according to a standard or known measurement. This ensures that when you zoom in on a specimen, the magnification levels are true to their specifications, giving you confidence in the accuracy of your observations and measurements. Regular calibration is particularly crucial when conducting quantitative research, where precision is paramount.

Alignment, on the other hand, is the harmonious arrangement of your microscope's optical components. Proper alignment ensures that light travels optimally through the lenses, resulting in images that are bright, sharp, and free from distortion. Misalignment, even by a small margin, can lead to blurry images or uneven illumination, which can significantly impact the quality of your observations.

So, how does one go about ensuring their microscope is both well-calibrated and aligned? Start with the manufacturer's guidelines, which should provide a roadmap for checking and adjusting calibration and alignment. Many microscopes come equipped with calibration slides or tools designed for this purpose. Using these tools, you can adjust the settings until they match the known standards, ensuring your microscope is accurately calibrated.

For alignment, the process often involves adjusting the condenser and ensuring that the light path is correctly centered. This may require you to tweak various knobs and settings, observing the changes in image quality until everything is perfectly aligned. It's a bit like solving a puzzle, where each piece must be in its right place for the entire picture to come into focus.

Remember, the goal of calibration and alignment isn't just about getting things "right" for the sake of it. These processes are fundamental to unlocking the full potential of your microscopy work, allowing you to explore the microscopic world with the utmost clarity and confidence. So, take the time to tune your instrument, and you'll be rewarded with a world of detail waiting to be discovered.

Common Problems and Their Solutions

Navigating the landscape of microscopy, you're bound to encounter a few bumps along the way. It's all part of the learning curve! Whether you're dealing with distorted images, poor contrast, or mechanical hiccups, each challenge presents an opportunity to deepen your understanding of your microscope's workings. Let's explore some of these common problems and their effective solutions, aiming to keep your journey through the microscopic world smooth and rewarding.

Distorted images often stem from improper lens alignment or issues with the specimen itself. If the world under your microscope seems a bit off-kilter, double-check that your specimen is mounted correctly and flat against the slide. A wrinkled or unevenly placed specimen can create odd visual effects. Next, ensure that your objective lenses are screwed in tightly and aligned properly; a lens that's even slightly askew can warp your view.

Poor contrast can be quite the conundrum, making it hard to distinguish your specimen from its background. This issue often has a simple fix: adjusting the diaphragm. The diaphragm controls the amount of light that reaches the specimen, and fine-tuning this setting can significantly improve contrast. Additionally, consider experimenting with different stains for your specimen, as these can enhance contrast and detail, bringing the invisible into sharp relief.

Mechanical failures, such as stuck knobs or a malfunctioning stage, can be frustrating but are usually within your power to remedy. A common cause of mechanical issues is debris or buildup in the moving parts. A gentle cleaning, or in some cases, a drop of lubricant can work wonders. However, if the problem persists, it's best to consult with a professional. Some issues, particularly those involving intricate internal mechanisms, require specialized knowledge and tools to fix.

Remember, each problem you solve not only keeps your research on track but also enriches your skills as a microscopy enthusiast. Embrace these challenges as part of your scientific adventure, knowing that with a bit of troubleshooting, you can overcome them and continue uncovering the marvels of the microscopic world.

Preventive Measures to Avoid Microscopy Pitfalls

Embarking on a microscopy journey without facing occasional hitches might seem like wishful thinking, but with preventive measures, you can significantly reduce the likelihood of running into common pitfalls. Think of your microscope as a valued team member in your research endeavors—one that thrives on attention and preventative care. Here's how you can keep your microscopy work on track while fostering a harmonious relationship with your equipment.

Regular cleaning cannot be emphasized enough. While we've touched on the importance of keeping your microscope clean, adopting a proactive stance towards this practice is key. Dust and grime are not just superficial nuisances; they can impair your microscope's performance over time. Establishing a routine that includes a quick clean before or after each use can work wonders in preserving image quality and functionality.

Another cornerstone of prevention is mindfulness in handling and usage. Microscopes are precise instruments, and as such, they demand a gentle touch and thoughtful handling. Ensure that every adjustment, whether it's the focus knob or the stage controls, is made with care. Avoid forcing any part of the microscope to move—if it resists, it's likely a sign that something needs your attention, possibly cleaning or realignment.

Routine checks go beyond cleaning; they're about staying ahead of potential issues. Take the time to periodically assess the alignment and calibration of your microscope. This doesn't just help maintain accuracy in your work but also preemptively addresses minor discrepancies before they morph into larger problems.

Lastly, fostering a culture of care among all who use the microscope reinforces the importance of preventive measures. Share insights and tips on maintenance and handling with peers. A collaborative approach to microscope care ensures that everyone benefits from a well-maintained instrument, enhancing the overall research experience.

By integrating these preventive measures into your microscopy routine, you're not just avoiding common pitfalls; you're ensuring that your pathway to discovery remains clear and unobstructed.

CONCLUSION

A Journey of Discovery

As we reach the end of "Microscopy for Beginners," I hope this book has ignited a spark of curiosity and wonder within you. Microscopy is not just a scientific tool; it is a gateway to a hidden world teeming with life, beauty, and complexity. Whether you are exploring the microscopic realm for academic, professional, or personal reasons, the skills and knowledge you have gained will serve you well on your journey of discovery.

Continuing Your Exploration

The world of microscopy is vast and ever-evolving, offering endless opportunities for exploration and learning. As you continue your journey, I encourage you to seek out new challenges, experiment with different techniques, and share your discoveries with others. Joining microscopy communities, attending workshops, and pursuing further education are great ways to deepen your understanding and connect with fellow enthusiasts.

Embracing the Wonder of Science

At its core, microscopy is about more than just magnifying objects; it is about expanding our perspective and understanding of the world. By embracing the wonder of science, we not only enrich our own lives but also contribute to the collective knowledge of humanity. Each observation, each discovery, brings us closer to unraveling the mysteries of the universe and inspires us to keep pushing the boundaries of what is possible.

Thank You

I would like to express my gratitude to you, the reader, for embarking on this journey with me. I hope this book has been a valuable resource and inspiration for your exploration of microscopy. Remember, the world beneath the lens is vast and full of surprises. Keep exploring, keep learning, and keep marveling at the wonders of the microscopic world.

Farewell, for Now

As we conclude our exploration of microscopy, I bid you farewell, but only for now. The world of science and discovery awaits, and I have no doubt that your journey will be filled with excitement, wonder, and endless possibilities. May your microscope reveal new worlds and your curiosity lead you to new discoveries. Thank you for joining me on this adventure. Until we meet again, happy exploring!

- Brian Rodgers

www.ingramcontent.com/pod-product-compliance
Lightning Source LLC
Chambersburg PA
CBHW071832210526
45479CB00001B/95